Electrical Generation and Distribution Systems and Power Quality Disturbances

Electrical Generation and Distribution Systems and Power Quality Disturbances

Editor

Isak Olsen

scitus academics

Electrical Generation and Distribution Systems and Power Quality Disturbances
Edited by **Isak Olsen**

ISBN: 978-1-68117-296-5
Library of Congress Control Number: 2016937481

SCITUS Academics LLC,
www.scitusacademics.com
Box No. 4766, 616 Corporate Way,
Suite 2, Valley Cottage,
NY 10989

Printed in United States of America on Acid Free Paper

Preface

Electricity generation is the first process in the delivery of electricity to consumers. The fundamental principles of electricity generation were discovered during the 1820s and early 1830s by the British scientist Michael Faraday. His basic method is still used today: electricity is generated by the movement of a loop of wire, or disc of copper between the poles of a magnet. Before Michael Faraday had discovered his famous law of electromagnetic induction, battery were the only source of electric power. After that, DC generator was developed, but it could produce only a few hundred volts of electric power and naturally this low voltage power could not transmitted efficiently to a large distance. In the latter half of eighteen centuries, AC electric power generation, transmission and distribution came into the picture. In an AC system, it became possible to step up voltage of electric power to desire level for efficient transmission to a long distance. After that 3-phase induction motor was developed which was much simpler in construction. Generation, transmission and distribution of AC power were much easier than DC power; hence very fast AC power system became the most popular means of electric power. Electricity is most often generated at a power station by electromechanical generators, primarily driven by heat engines fueled by chemical combustion or nuclear fission but also by other means such as the kinetic energy of flowing water and wind. Other energy sources include solar photovoltaics and geothermal power and electrochemical batteries. Additionally, the presence of a static converter as output interface of the generating plants introduces voltage and current harmonics into the electrical system that negatively affect system power quality. By integrating distributed power generation systems closed to the loads in the electric grid, we can eliminate the need to transfer energy over long distances through the electric grid. This book, Electrical Generation and Distribution Systems and Power Quality Disturbances, reveals the different power generation and distribution systems with an analysis of some types of existing disturbances and a study of different industrial applications such as battery charges.

Table of Contents

CHAPTER 1

Integrated Coordinated Optimization Control of Automatic Generation Control and Automatic Voltage Control in Regional Power Grids

Qiu-Yu Lu [1], Wei Hu [1,*], Le Zheng [1], Yong Min [1], Miao Li [2], Xiao-Ping Li [2], Wei-Chun Ge [3] and Zhi-Ming Wang [3]

[1] State Key Lab of Power Systems, Department of Electrical Engineering, Tsinghua University, Beijing 100084, China

[2] Hubei Electrical Power Co. Ltd., Wuhan, Hubei 430077, China

[3] Liaoning Electrical Power Co. Ltd., Shenyang, Liaoning 110006, China

ABSTRACT

Automatic Generation Control (AGC) and Automatic Voltage Control (AVC) are key approaches to frequency and voltage regulation in power systems. However, based on the assumption of decoupling of active and reactive power control, the existing AGC and AVC systems work independently without any coordination. In this paper, a concept and method of hybrid control is introduced to set up an Integrated Coordinated Optimization Control (ICOC) system for AGC and AVC.

Concerning the diversity of control devices and the characteristics of discrete control interaction with a continuously operating power system, the ICOC system is designed in a hierarchical structure and driven by security, quality and economic events, consequently reducing optimization complexity and realizing multi-target quasi-optimization. In addition, an innovative model of Loss Minimization Control (LMC) taking into consideration active and reactive power regulation is proposed to achieve a substantial reduction in network losses and a cross iterative method for AGC and AVC instructions is also presented to decrease negative interference between control systems. The ICOC system has already been put into practice in some provincial regional power grids in China. Open-looping operation tests have proved the validity of the presented control strategies.

KEYWORDS

Integrated Coordinated Optimization Control (ICOC); Automatic Generation Control (AGC); Automatic Voltage Control (AVC); discrete event-driven; hierarchical control; cross iterative calculation

1. INTRODUCTION

Automatic Generation Control (AGC) and Automatic Voltage Control (AVC) systems are crucial parts in a modern Energy Management System (EMS). Active power dispatch and frequency regulation in China consists of three parts: Economic Dispatch Control (EDC) ahead of day, Load Frequency Control (LFC) and Primary Frequency Regulation (PFR) for real time. EDC is formulated to satisfy the hourly load demand based on load prediction ahead of day; LFC is responsible for frequency volatility with cycles ranging from 10 seconds to several minutes, while PFR deals with the real time balance of generation and load [1]. LFC is implemented by an AGC system, and in recent years integrated EDC/AGC systems have been widely used and are regarded as the standard AGC system. On the other hand, since several blackouts took place in the world's large power grids due to voltage instability and collapse, voltage stability and control has attracted worldwide attention among researchers and industrial professionals, with a number of significant achievements being made [2,3]. One of them is the hierarchical voltage control model, which can effectively prevent voltage collapse through optimization of reactive power distribution. The hierarchical model has laid the foundation of the AVC system. It should be noted that due to the physically weak coupling between P and Q [4], the existing AGC and AVC systems work independently without interaction or coordination. However, the separate optimization method is unreasonable in the following aspects:

- Active and reactive power regulation are both constrained by the generator capacity limit, so separate optimization is not able to guarantee that the generator is operating in the secure zone;
- Line loss is part of the active power output, but the conventional Loss Minimization Control (LMC) model only considers reactive power optimization. From the perspective of economical operation, the active and reactive optimization should not be separated;
- Active power/frequency control and reactive power/voltage control are not completely decoupled even under normal conditions, for example, frequent fluctuation of active power affects the voltage quality; interactions between AGC and AVC control commands result in weakened control effects and reciprocal regulation. Furthermore, the weak coupling relation no longer exists in overload conditions and the independent control model may cause security problems.
- Recently, with the increasing penetration of wind power energy, its impact on power system operations cannot be neglected. Wind generation operation requires a large amount of reactive power support and flexible frequency regulation ability, hence putting higher demands on AGC and AVC systems, especially for the cascading trip-off accident tangling with active and reactive power control at the same time [5].

With the development of information and automatic control technologies, coordinated control of active and reactive power is recognized as a major trend. This is also in line with the integration concept in smart grids, with all smart devices and control systems taken into consideration [6].

To date, research on coordination of active and reactive power is largely unreported. In [7] a new dynamic power flow algorithm was proposed, which helps avoid unreasonable power flow distribution due to bus type definition through joint adjustment of active and reactive power. Concerning economically coordinated dispatch, in [8], an active/reactive coordinated optimization model is set up to allocate network losses to each generator under the principle of minimizing total generation cost, but in dealing with conflicting inequality constraints, the converting bus and branch type methods proposed in this paper may cause frequent transformation and affect convergence of the algorithm. Other research is based on prediction control. Through ultra-short load forecasting, active/reactive power flow can be optimized to minimize network losses and maximize loading margin [9,10], yet real-time coordinated control is not involved. Currently all relevant research is largely in the theoretical exploration phase, without reports about AGC and AVC coordination in the engineering field.

The concept and control strategy of an Integrated Coordinated Optimization Control (ICOC) system for AGC and AVC is introduced in this paper from an engineering perspective. The ICOC system is designed in a hierarchical structure with respective targets and control objects for each layer. In the top layer, three types of discrete events, regarding security, quality and economic events, are defined to drive the responding joint optimization block. The hierarchical control and event-driven mechanism significantly reduce the complexity of the multi-target optimization problem concerning diverse control devices. Meanwhile an innovative LMC model adopting both active and reactive power regulation is proposed to substantially reduce network losses. In addition, to minimize the negative interaction between AGC and AVC commands and improve operation quality, an instruction cross iterative method is also presented. Based on the control framework and strategies mentioned, practical ICOC systems have already been established and open-looping control tests conducted in some regional power grids in China. Pre-operation of ICOC proves the validity of the presented control strategies. The remainder of this paper is structured as follows: Section 2 describes the system structure and control model of the proposed ICOC system. Section 3 gives the detailed design proposal of a practical ICOC system according to the technical conditions of a provincial power grid. Section 4 demonstrates the single section and continuous simulation results of the ICOC system, and analysis of the advantages of coordinated control. Section 5 concludes this paper by summarizing the unique characteristics and significance of ICOC system.

2. SYSTEM STRUCTURE AND CONTROL MODEL OF ICOC

The ICOC system adopts the discrete event-driven pattern and hierarchical structure from the concept of hybrid control theory [11]. ICOC takes into account the coupling relation between active power/frequency and reactive power/voltage, integrating the original optimal indices and modifying the conventional algorithms. By making full use of control resources, the ICOC system is able to improve system stability, operation quality and economy. The hierarchical structure of the ICOC system is shown in Figure 1.

The ICOC system consists of three layers: Top Layer for Management and Decision, Middle Layer for Coordination and Bottom Layer for Execution. Each layer is designed with its own responsibilities, including receiving control targets from the upper layer, performing optimization calculations, as well as forming instructions and sending them to the lower layer. Overall regulation in a hierarchical approach is thus carried out, helping to achieve an integrated optimal target as expected.

2.1. Top Layer for Management and Decision

The Top Layer for Management and Decision, which monitors the operational status of the power system through Supervisory Control and Data Acquisition (SCADA) and Wide Area Measurement System (WAMS), analyzes the security, quality and economic status to judge whether the system is under optimal or quasi-optimal conditions. Once any index exceeds its limits, an event will be formed to drive a responding control block, which then works out commands and sends them to the Middle Layer.

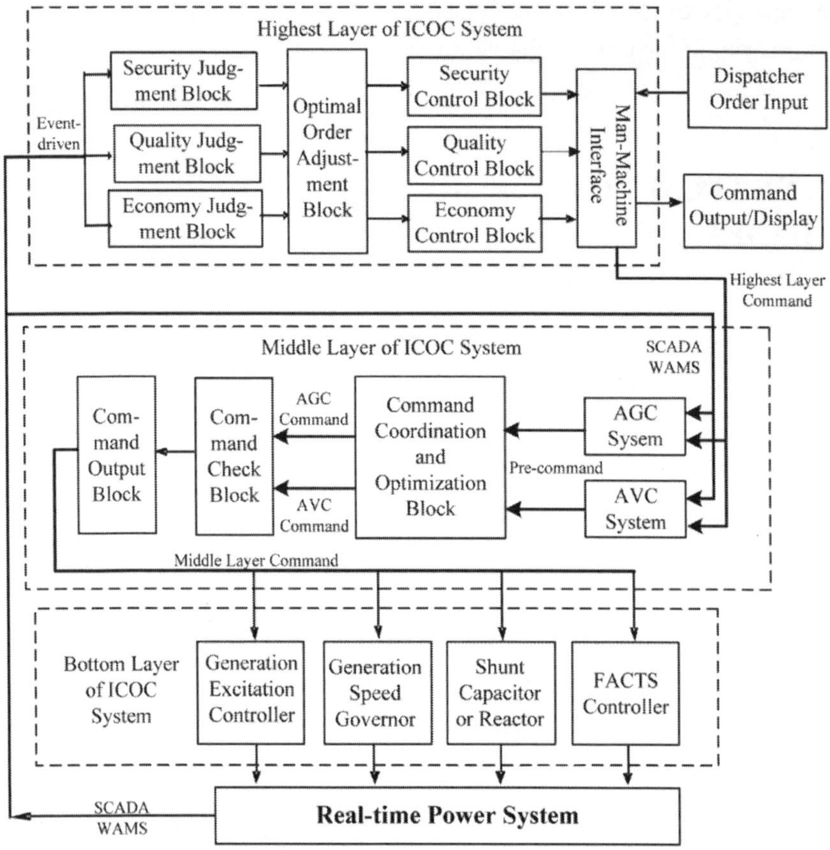

Figure 1. Hierarchical structure of the ICOC system.

2.1.1. Economic Event and Control Model

The economic event monitors the network losses and can be defined as follows:

$$E_{eco} = \begin{cases} C_{econ} & \text{if } P_{loss}[k] > P_{loss-eco} \\ C_{enon} & \text{if } P_{loss}[k] \le P_{loss-eco} \end{cases} \quad (1)$$

where $E_{eco} = \{C_{econ}, C_{enon}\}$ stands for the output of economy judgment block, C_{econ} and C_{enon} represent the need and no need for regulation, respectively. $P_{loss\text{-}eco}$ refers to the maximum network losses permitted, and $P_{loss}[k]$ is the actual network loss at control interval k.

The economic control block is driven by a discrete event from the judgment block. If the event received is C_{enon}, the output command will stay unchanged; if the event is C_{econ}, then economic analysis will be conducted according to the current condition, and output command will be updated and sent to the Middle Layer. In the decision-making process, the following optimal problem is solved:

$$\begin{aligned}
&\min J\left(P_g, V_g\right) = \min f(P_g) + \eta P_{loss}(P_g, V_g) \\
&s.t. \quad P_i = V_i \sum V_j (G_{ij} \cos\theta_{ij} + B_{ij} \sin\theta_{ij}) \\
&\qquad\;\; Q_i = V_i \sum V_j (G_{ij} \sin\theta_{ij} - B_{ij} \cos\theta_{ij}) \\
&\qquad\;\; \underline{V}_i \le V_i \le \overline{V}_i \\
&\qquad\;\; \underline{V}_g \le V_g \le \overline{V}_g \\
&\qquad\;\; \underline{P}_g \le P_g \le \overline{P}_g
\end{aligned} \tag{2}$$

Equation (2) is an innovative LMC model, taking advantage of active power regulation, which plays a key role in power flow distribution. The optimal targets in (2) are the network losses P_{loss} and active power regulation cost f (due to the fact that reactive power regulation is a basic service free of charge, while active power regulation needs to meet clear cost criteria). On-grid price η ($/MW) is adopted to unify the two elements into currency unit ($), thus enabling integrated energy-savings in both power plants and power grids. The base value of AGC units P_g and terminal voltage of AVC units V_g make up the control variables, and system network functions are the equality constraints while tolerable bus voltage range and control variable limits form the inequality constraints. It should be noted that the AGC output is the sum of power base point and frequency adjustment value [12]. In a conventional control model, the base point is determined by the Unit Commitment model in EDC while the adjustment value is responsible for

Area Control Error (ACE) in LFC. In this paper, the base point is included in the real-time LMC model, which is feasible as the AGC cycle matches the LMC control time scale. However, in order to guarantee normal ACE regulation, P_g is supposed to stay within the AGC adjustment capacity with sufficient buffer. An interior point method [13] is used to solve the large scale optimal power flow problem as above, and results (base point of AGC P_g and pilot bus voltage V_P^{ref} which stands for area voltage level) are sent to the Middle Layer.

2.1.2. Security Event and Control Model

Among the various indices applicable to evaluate voltage stability, the minimum amplitude eigenvalue (MAE) of the Jacobian Matrix in power flow calculation λ_{min} is selected as the monitoring index of security events. Other indices are also feasible for this block with appropriate modification, but are not illustrated in this paper. The security event is defined as follows:

$$E_{sec} = \begin{cases} C_{scon} & \text{if } \lambda_{\min}[k] < \lambda_{sco} \\ C_{snon} & \text{if } \lambda_{\min}[k] \geq \lambda_{sco} \end{cases} \quad (3)$$

where $E_{sec} = \{C_{scon}, C_{snon}\}$ stands for the output of the security judgment block, C_{econ} and C_{enon} represent the need and no need for voltage stability control, respectively. λ_{sco} is the pre-set MAE value for system security, and λ_{min} $[k]$ is the actual value at control interval k. If needed, the security control block will be driven to perform a stability analysis and update the commands; if not, the output will stay unchanged.

The security control block analyses the system's stability level according to λ_{min}, and works out the security weight coefficient ω which is related to multi-target optimization of AVC in the Middle Layer. The security weight coefficient with respect to network structure, typical operating conditions and reactive power reservation levels, can be referred to in the strategy schedule made by a series of off-line calculations. The most up-to-date security weight coefficient is feasible for the actual condition, providing benefits in emergency control situations.

2.1.3. Quality Event and Control Model

Fundamentally, AGC and AVC are designed to maintain the system's frequency and tie-line power flow and voltage in the accepted range and ensure the system's operation quality. Therefore, ACE and pilot bus voltage are adopted as monitoring indices for quality event.

The quality event is defined as follows:

$$E_{qua} = \begin{cases} C_{qcon} & \text{if } |\Delta V_P[k]| > \Delta V_{qco} \quad \text{or} \quad ACE[k] > ACE_{qco} \\ C_{qnon} & \text{if } |\Delta V_P[k]| \le \Delta V_{qco} \quad \text{and} \quad ACE[k] \le ACE_{qco} \end{cases} \tag{4}$$

where $E_{qua} = \{C_{qcon}, C_{qnon}\}$ stands for the output of quality judgment block, C_{econ} and C_{enon} represent the need and no need for quality control respectively. ΔV_{qco} is the maximum permitted voltage deviation and ACE_{qco} is the maximum permitted ACE. $\Delta V_P[k]$ is the deviation between the actual pilot bus voltage $V_P[k]$ and the reference voltage $V_P^{ref}[k]$ given by the economic block at control interval k ($\Delta V_P[k] = V_P^{ref}[k] - V_P[k]$). $ACE[k]$ is the measured value at control interval k.

If the output event is C_{qnon}, the command will stay unchanged; if the output event is C_{qcon}, the quality control block will figure out the total active power adjustment value through $\Delta P = ACE[k]$ and pilot bus voltage adjustment value $\Delta V_P[k]$ for each control area and send them to the Middle Layer.

2.1.4. Optimal Order Adjustment of Discrete Events

An optimal order adjustment module is set up to deal with the situation when two or three types of events take place simultaneously. In line with the priority order of power system's control targets, security events are given top priority, followed by quality events and then economic events. Each time the top priority event drives the corresponding control module and forms commands with the aim to eliminate the event and sends them to the Middle Layer. Therefore only one type of event is eliminated in every control cycle, and the system status changes as shown in Figure 2. The optimal order adjustment module improves the control system efficiency.

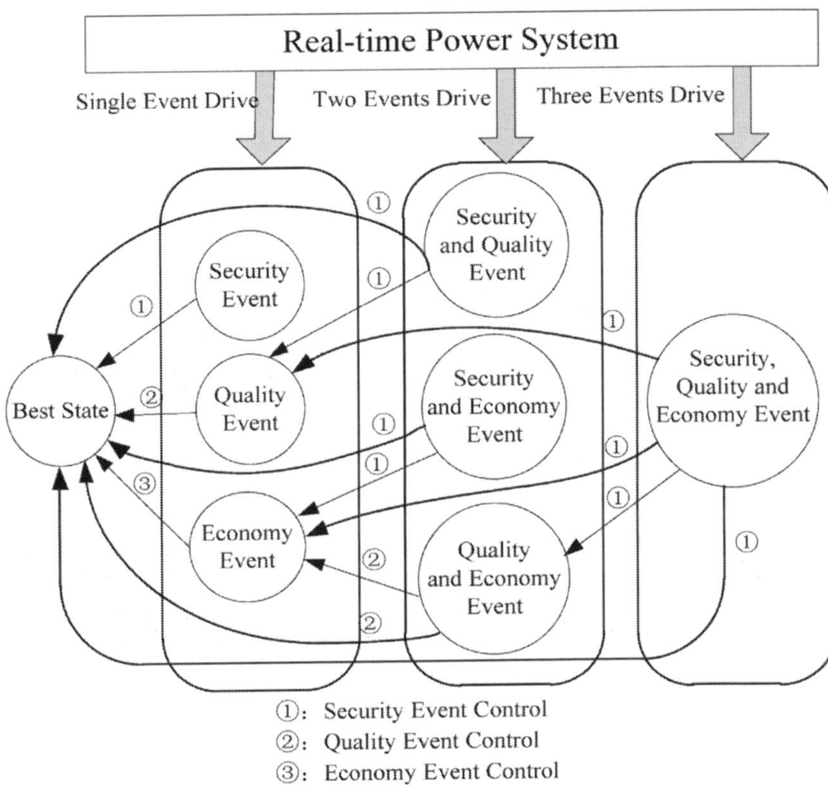

Figure 2. ICOC status transition diagram.

The event-driven pattern abandons the conventional fixed-time mechanism resulting in untimely control. The Top Layer, driven by discrete events, takes measurements to eliminate the events and guarantees the following regulations at optimal trajectory with efficiency and timeliness. Moreover, instead of solving multi-target optimal problems, the event-driven pattern turns to pursue quasi-optimization. The system deals with single type of event at each interval but the output commands of un-triggered events remain unchanged, thus the solvability is guaranteed and coordinated optimal control of diverse targets can be achieved. Therefore, the ICOC system can significantly reduce the network losses and meanwhile ensure the dynamic quality and stability of the power system.

2.2. Middle Layer for Coordination

In the Middle Layer for Coordination, conventional control strategies are modified to realize coordination of different optimal control targets and devices. Furthermore, as the interaction between AGC and AVC commands is taken into full consideration, the instruction cross iterative method is proposed to revise the pre-commands, so as to prevent reciprocal regulation by the two control systems. After iteration, the coordinated commands of AGC and AVC systems are sent to the Bottom Layer.

2.2.1. AGC Control Strategy

As mentioned above, the AGC output consists of two parts: base point and adjustment value. In this paper, base point is decided by the economic control block of the Top Layer, while the total active power adjustment value is provided by the quality control block. The AGC strategy is supposed to allocate the total ACE power to each AGC unit and form the AGC pre-command together with the base point. Conventional AGC power allocation is based on AGC capacity or economic factors. However, in the context of a market economy, the impact of AGC allocation on network losses needs to be taken into consideration due to the wide-area distribution of AGC units. In this paper the network losses sensitivity with respect to AGC output [14] is applied to correct the economical allocation factor, and form the comprehensive economical index. The index guides the allocation of total frequency adjustment power as follows:

$$\begin{aligned} &\min \sum_i (a_i + \frac{dP_L}{dP_i}) \times \Delta P_i \\ &s.t. \sum_i \Delta P_i = \Delta P \\ &\quad | \Delta P_i | \leq \Delta P_{i\max} \\ &\quad P_{i\min} \leq P_i + \Delta P_i \leq P_{i\max} \end{aligned} \tag{5}$$

where a_i is regular economical allocation factor of generator *i* from the Economic Dispatch Control; dP_L/dP_i refers to the network losses sensitivity with respect to the active power output of generator *i*; thus the sum of a_i and dP_L/dP_i is the new economical factor. ΔP_i represents the adjustment power of generator *i*, ΔP stands for the total adjustment value from the Top Layer, ΔP_{imax} is the maximum ramp rate of generator *i*, while P_{imin} and P_{imax} are the minimum and maximum power output, respectively. Therefore the economic effect of active power regulation is given full play under the new allocation principle in Middle Layer and the new LMC model in Top Layer. Consequently, the pre-command of AGC is given by Equation (6):

$$\Delta P_{AGCi} = P_{gi} + \Delta P_i \tag{6}$$

2.2.2. AVC Control Strategy

Ensuring sufficient reservation for dynamic reactive power is an effective approach to enhance the voltage stability of a power system, but the conventional method based on voltage sensitivity leads to imbalanced distribution of reactive power, and generators with greater sensitivities may reach their limits, resulting in lower voltage stability margins, or even significant voltage drops or voltage collapse in the case of heavy flow or lack of reactive power. Therefore some researchers have proposed the Coordinated Voltage Control theory [15,16]. Different stability indices guide the coordination between economical operation and security control.

As illustrated in reference [17], balanced generator reactive power distribution contributes to voltage stability. Therefore regional unified of reactive power distribution Q_{lev} is adopted as the voltage security index, as defined in (7):

$$Q_{lev} = \Sigma(k_i - \bar{k})^2, where\ k_i = \frac{Q_i - Q_{i\min}}{Q_{i\max} - Q_{i\min}} \tag{7}$$

where Q_i is reactive power of generator i, while Q_{imin} and Q_{max} are the minimum and maximum reactive power, respectively. Thus k_i refers to the percentage of reactive power output of generator i, and i refers to the average of k_i in the same control region. Therefore, the AVC control strategy can be described by (8):

$$\begin{aligned} &\min \Sigma \Delta Q_i^2 + \omega Q_{lev} \\ &s.t. \mid \Delta V_p + C\Delta Q \mid < \varepsilon \\ &\quad V_{i\min} \le V_i + C_i \Delta Q_i \le V_{i\max} \\ &\quad Q_{i\min} \le Q_i \le Q_{i\max} \end{aligned} \tag{8}$$

where ΔQ_i is reactive power adjustment value of generator i, C is voltage sensitivity vector with respect to the reactive power of generators, C_i is the terminal voltage sensitivity with respect to the reactive power of generator i. The optimal targets in (8) include reactive power adjustment and regional unified index of reactive power distribution, which stands for a combination of economical operation and system security. The inequality equations as the constraints define the acceptable voltage deviation of pilot bus voltage and the limits of reactive power and terminal voltage. In addition, the control target of pilot bus voltage deviation value ΔV_P and security weight coefficient ω are received from the Top Layer.

2.2.3. Iteration of AGC and AVC Pre-Commands

The change of AGC active power leads to redistribution of power flow, thus affecting the pilot bus voltage. Meanwhile, since the changes in network losses after AGC and AVC regulation are not taken into consideration during the ACE allocation, control inaccuracy is unavoidable. In particular, in the case of remarkable total frequency adjustment power, the AGC system sends frequent commands, posing a negative impact on generators and voltages. As the changes in network losses and control inaccuracy in pilot bus voltage reflect the interactive impact between AGC and AVC systems, the two elements are adopted as monitoring indices, which builds an effective communication bridge between the AGC and AVC systems. The instruction cross iterative method is applied to minimize the negative effects in this paper. In the interaction process, optimization of active and reactive power is decoupled with different goals (ΔP_{loss} and ΔV_P). Due to the impact of active and reactive power control being associated with adjustment, the interactive impact between the two systems gradually shrinks with the declining amount of regulation commands, and the monitoring indices eventually converge to the pre-set threshold. The iterative flow is shown in Figure 3.

By minimizing the indices in the iterative process, the ICOC system can improve the dynamic quality of frequency and voltage, and realize delicate regulation of active and reactive power. Moreover, as the cross iterative algorithm decouples the optimization of AGC and AVC, only slight modification of original control systems is necessary, making the iterative flow practical in installation.

2.3. The Bottom Layer for Execution

The Bottom Layer for Execution consists of various active and reactive regulation devices in plants and substations. Controllers receive coordinated commands from the Middle Layer, change the operation point and finally realize the coordination optimization control of AGC and AVC, enabling economic, quality and security operation of power system.

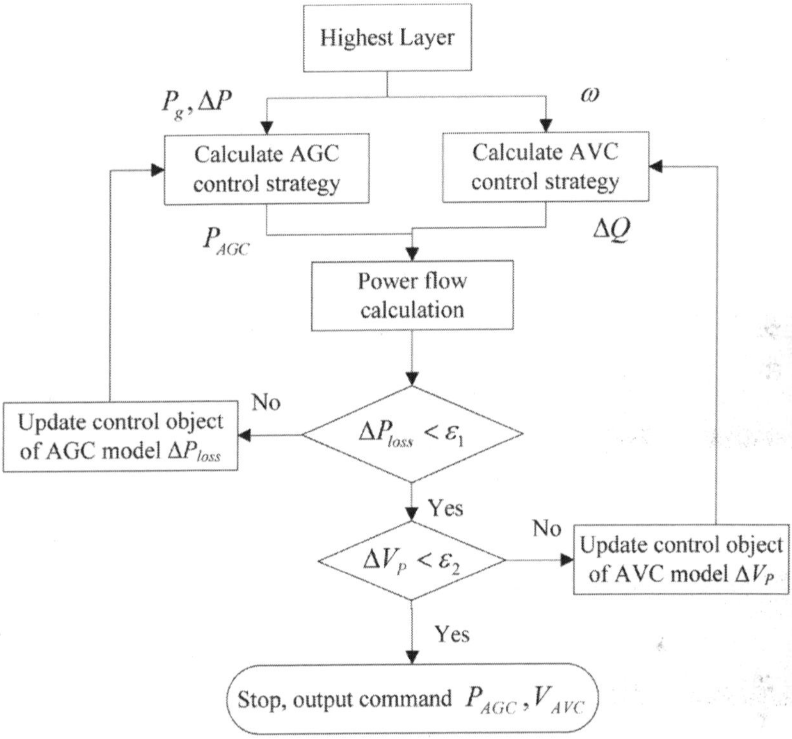

Figure 3. Iterative flow of AGC and AVC pre-commands in Middle Layer.

3. DESIGN AND IMPLEMENTATION OF ICOC SYSTEM

Based on the presented concept of coordinated optimization control, a practical ICOC system is designed in this section according to the existing hardware and software specifications in China's provincial regional power grids. The proposal includes system configuration of hardware and software, system interface design, core functional module development, as well as system reliability and safety assurance.

3.1. Hardware Configuration

The hardware system of the ICOC is configured with two Linux servers and two workstations installed in the dispatch centers of provincial power grids. The application server is responsible for the analysis and calculation of various application function modules in the platform, while the database server provides specialized storage for related data and application information using the professional Oracle database for the ICOC system. Connected with servers, workstations provide an interface operation and debugging environment. The ICOC system is linked to the Dispatching Automatic System through a special network channel to enhance the data transfer rate. As a result, the ICOC system can easily read real-time data from dispatch systems and automatically stayed synchronized with the network models of the dispatch platform. Additionally, the hardware firewall and physical insulation equipment are installed to ensure the secure separation among sub-systems and enhance the safety of the system. The hardware configuration and its interfaces with other equipment are shown in Figure 4(a).

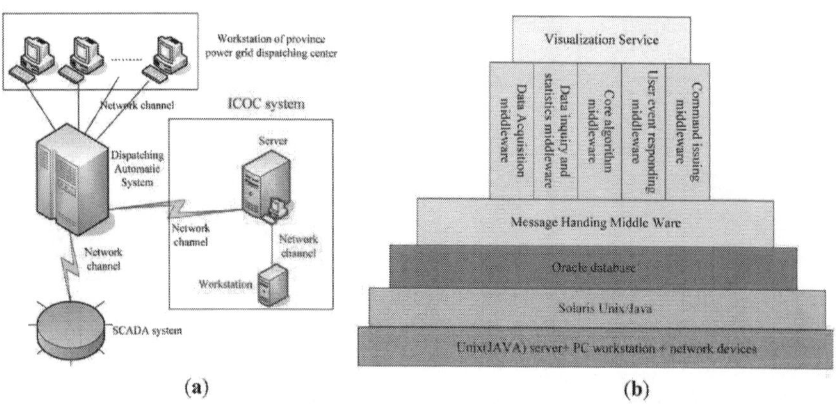

Figure 4. ICOC hardware configuration and software architecture.

3.2. Software Architecture

The hierarchical component architecture is adopted in the software system of the ICOC. Four layers are constructed based on the hardware platform, including Operating System Layer, Database Layer, Middle Ware Layer and Visualization Service Layer. Software functions (including data acquisition, data inquiry and statistics, core algorithm, user event response and instruction-sending) are integrated in the middle ware, interacting with each other through Message Handling Middle Ware, and fulfill ICOC's main functions. The integrated software architecture is shown in Figure 4(b).

3.3. Design Details of Core Algorithm

The core algorithm is the most crucial and complicated module in the ICOC System and it consists of three sub-modules: advanced application module, basic module and interface model, as in Figure 5.

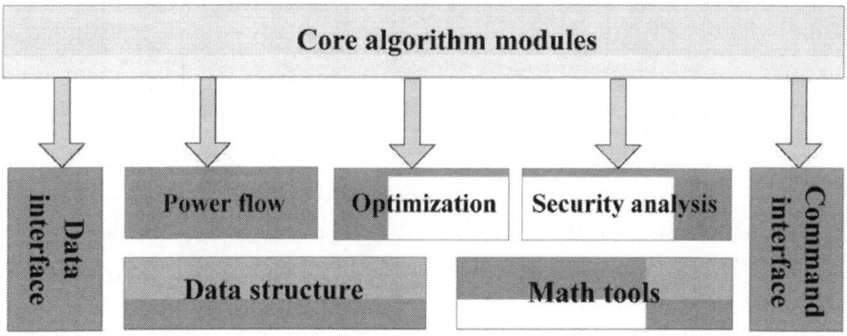

Figure 5. Core calculation and analyses modules in ICOC.

- Advanced application module: The advanced application module consists of three modules including power flow calculation, optimization and security analysis. Power flow module is designed for basic flow calculations, including data error correction, network structure equivalent, network loss calculations and sensitivity calculations. The optimization module is designed for solving the optimal power flow model, AGC and AVC command optimization and iteration. The security analysis module is in charge of assessing voltage security in static status, calculating voltage security indices and conducting AVC secure weight coefficient schedule.

- Basic module: The basic module consists of the data structure module and mathematical tools. The data structure module manages various data and conducts data transfer among different modules. The mathematics library provides a wealth of mathematical tools and optimization algorithms for the three advanced application modules.
- Interface module: The interface module includes the data interface module and command interface module, which interact with the Message Handling Middle Ware, deliver data and send instructions.

4. APPLICATION OF THE ICOC SYSTEM IN CHINA'S REGIONAL POWER GRIDS

The proposed ICOC system is applied in the power grid of central China's Hubei Province. Two-thirds of the system load is distributed in the province's southeast region centered on the provincial capital Wuhan, while two-thirds of the power plants are located in the province's west region. As a result, large capacity power transmission over long distances in the power grid is required, posing difficulties for operation and control. Moreover, the fact that thermal generators represent the majority and flexible hydropower generators are inadequate, has restrained the regulation quality of AGC, and AVC control is applicable up to 220 kV level due to the limits of duties and assignment in the dispatch system. Under such a situation, coordinating AGC and AVC is considered the most effective way to enrich regulation approaches and improve the operation quality of Hubei power grid. Static and dynamic simulations were conducted in the ICOC system as follows.

4.1. Static Simulation

Static simulations consist of two parts: single section simulation and continuous simulation, which involved 16 AVC plants and AGC generators accounting for 10% of the total installation capacity.

4.1.1. Single Section Simulation

Actual historical power flow section on September 20th, 2010 was used for single section simulation. Three control schemes were conducted in comparison as follows:

- Scheme 1: only AGC control, and total regulation power is allocated under the equal capacity principle;
- Scheme 2: AGC and AVC systems work independently, and AVC optimal target is to minimize the volume of reactive power regulation;
- Scheme 3: AGC and AVC adopt the coordinated optimization method proposed in this paper.

The comparisons regarding economic, security and quality indices under the three schemes are depicted in Table 1.

- Economic index comparison Network loss reduction in Scheme 2 was not significant compared to the original status of Scheme 1 due to the limited reactive power regulation methods. Meanwhile, in Scheme 3 noticeable reductions in network losses were achieved as a result of coordination of active and reactive power, significantly improving the economical operation of the power grid.
- Security index comparison Uniformity of reactive power index Q_{lev} in Scheme 3 is the lowest, indicating that the active power distribution is most balanced under coordinated control. Statistics about generator distribution regarding reactive power output in different range are shown as in Figure 6, in which there are 15% and 6% generators in Schemes 1 and 2, respectively, close to the reactive power upper limit (more than 80%) or lower limit (less than 20%).

However in Scheme 3, no generators' reactive power was close to the limits and the reactive power margin is the largest. Therefore in case of system disturbance, dynamic reactive power can quickly provide support to maintain voltage stability. Meanwhile, comparison about security index λ_{min} also proves the static voltage stability in Scheme 3 is best among all.

- Quality index comparison The quality indices of active and reactive power (ACE and pilot bus voltage) in Scheme 3 all performed better than in Schemes 1 and 2, which indicated the cross iteration method can avoid reciprocal regulation of AGC and AVC systems and improve operation quality.

Table 1. Indices comparison of three schemes in single section simulation.

	Index	Scheme1	Scheme2	Scheme3
Economy	Network losses/MW	466.9	464.5	438.2
Security	Q_{lev} sum of whole grid	2.522	1.243	0.786
	λ_{min}	0.7488	0.7574	0.7587
Quality	ACE/MW	−1.60	0.72	−0.07
	Average voltage of pilot buss/p.u.	0.9963	1.004	1.006

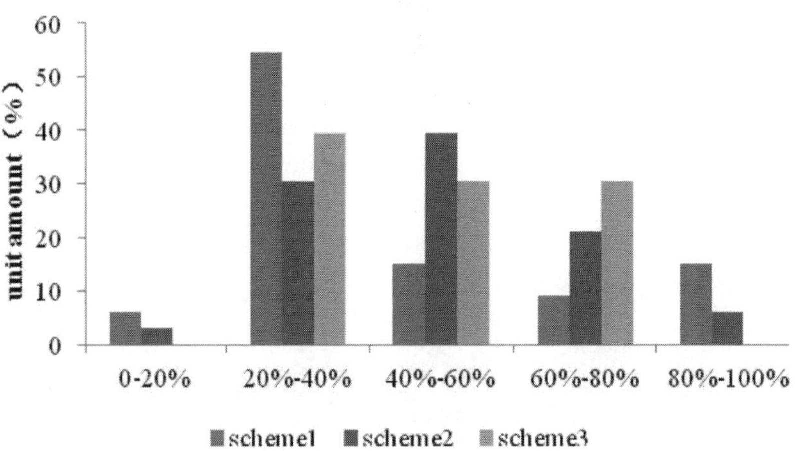

Figure 6. Generator distribution regarding reactive power output in different range.

4.1.2. Continuous Simulation

Continuous simulations were conducted with historical data during 8:30 A.M. September 20th and 8:30 A.M. September 21st, 2010. The control interval was 5 minutes. Comparison and analysis were made among the three schemes as mentioned earlier. (a) shows that the coordinated scheme proposed in this paper can effectively reduce network losses, the rate of which dropped by 0.08% on average during the simulation period. That means a potential of approximately 720 million kWh of electricity worth about 20 million RMB saved annually in the provincial power grid, and a reduction of 90,000 tons of carbon dioxide emissions at the same time. The ICOC system can greatly improve the economical operation of power systems as well as bring about considerable social and environment benefits. Moreover, the improvement in security and quality indices of coordinated scheme were demostrated in Figure 7(b–d). All the simulation results verify the validity of the proposed control method.

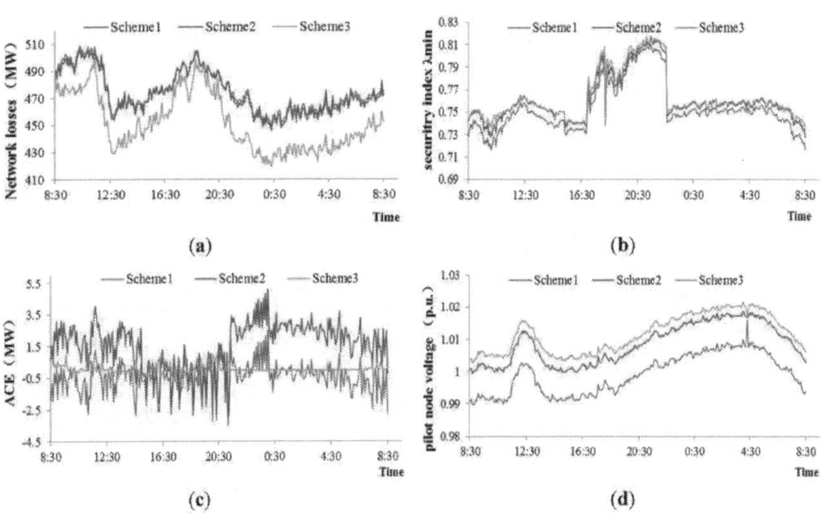

Figure 7. **(a)** Losses curve comparison of three schemes; **(b)** Security index λ_{min}comparison of three schemes; **(c)** ACE curve comparison of three schemes; **(d)** Pilot bus voltage comparison of three schemes.

4.2. Dynamic Simulation

In order to demonstrate the dynamic regulation process of the ICOC system and its effect on system performance and stability, active and reactive load was increased slowly in proportion on the weak nodes in the Wuhan area. The minimum control interval is 10 seconds. Three AVC plants and nine AGC plants participated in this control, and the maximum change step for AVC unit reference and AGC power are relatively 0.02 p.u. and 30 MW. Three control schemes as Section 4.1 were conducted, and the pilot node voltage curve was shown in Figure 8.

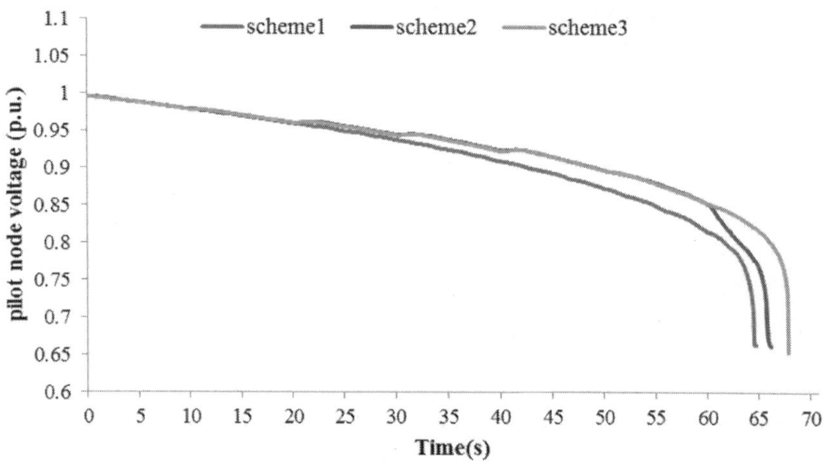

Figure 8. Pilot node voltage curves of three schemes.

It can be draw from Figure 8 that Scheme 3 obtained the maximum stability margin. On the other hand, in Scheme 2 voltage dropped suddenly at 60 seconds and thus accelerated the voltage collapse. Figure 9 demonstrates the regulation process of the three AVC plants and explains the reason for the voltage drop in Scheme 2. Scheme 2 was targeted at minimizing the regulation amount, therefore the AVC plant 2, which is most sensitive to pilot node voltage, was driven to devote more reactive power in the regulation process and reached its limit at 53 seconds. Due to the action of the over excitation limiter, its reactive power was forced to decrease, as shown in Figure 9(f), and this resulted in the acceleration of voltage collapse. Scheme 3 coordinated the reactive power regulation amount and uniformity, thus decreased the ascend speed of plant 2 and drove plants 1 and 3 to make more contribution instead. Therefore, it maintained the reactive power reservation in the weak area.

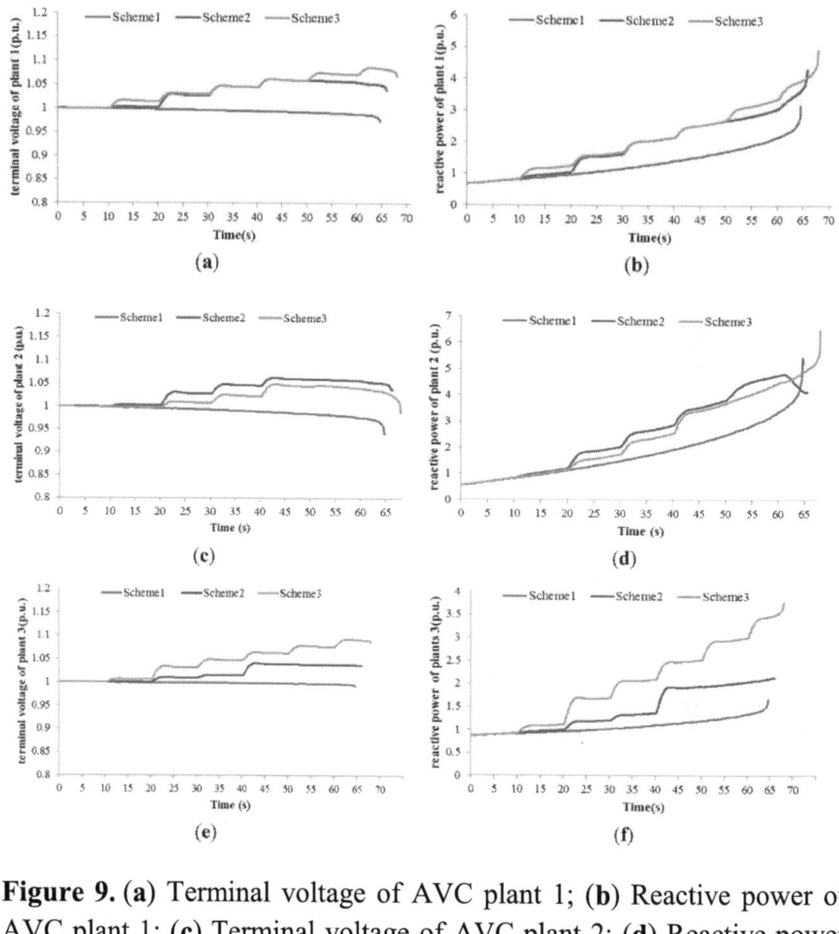

Figure 9. (**a**) Terminal voltage of AVC plant 1; (**b**) Reactive power of AVC plant 1; (**c**) Terminal voltage of AVC plant 2; (**d**) Reactive power of AVC plant 2; (**e**) Terminal voltage of AVC plant 3; (**f**) Reactive power of AVC plant 3.

The active power of two AGC units was shown in Figure 10 as representative, other AGC curves were not given here due to paper length limit. Both of the AGC strategies respond correctly and maintain the quality of frequency and tie line power. The ICOC system gave priority to the AGC units in heavy load area, thus reducing the long distance power transmission and improving the voltage stability.

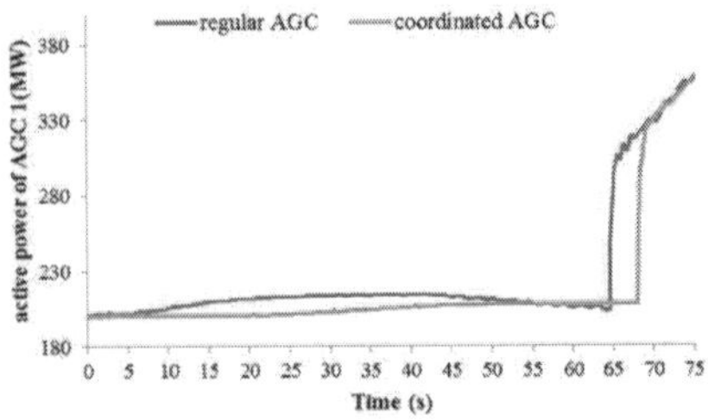

(a)

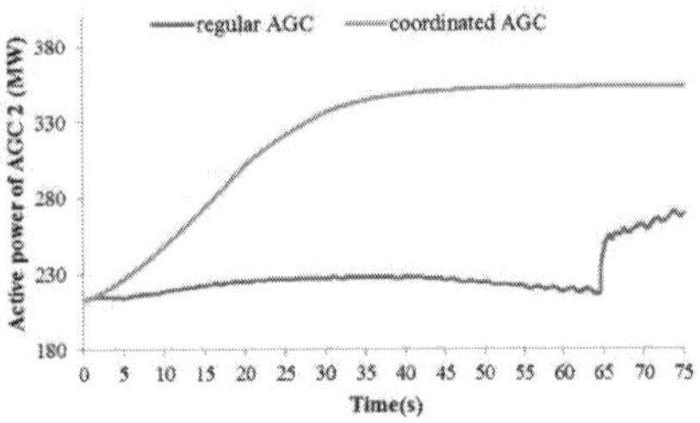

(b)

Figure 10. (**a**) Active Power of G1 (**b**) Active Power of G2.

5. CONCLUSIONS

Based on the coupling relation of active and reactive power control, event-driven mechanism and hierarchical control theory, an ICOC sytem of AGC and AVC is proposed in this paper, with the following characteristics:

- **Event-driven mechanism.** Well-directed control measures are taken under an event-driven mechanism for the system's security, quality and economic events, avoiding the unnecessary or untimely control in the time-fixed control and greatly improving regulation effects.
- **Innovative LMC model integrating active and reactive power regulation.** The AGC system is adopted in the LMC model, including the base point adjustment in optimal flow model of the Top Layer and allocation principle modification by network loss sensitivity in the Middle Layer. Thus the innovative LMC model gives full play to various active and reactive control measures and

conducts smooth coordination among them to minimize network losses and enhance economical operation.

- **Coordinated AVC control module.** The uniformity index of reactive power distribution is proposed in this paper, and its weight is self-adaptive for real-time operation under the control of security events. Coordinating the considerations for economical operation and security in voltage control, the module is effective in preventing voltage collapse and widespread blackouts.
- **AGC and AVC command cross iterative method.** AGC and AVC commands are modified according to interactions between the two systems, which decouple the optimization of AGC and AVC for easier application in current control systems, yet at the same time builds an effective communication bridge between the two systems. The iteractive method enables better control quality and delicate regulation for active and reactive power.

The coodinated control method was applied in the power grid of central China's Hubei Province. An ICOC system was designed and installed based on the grid's existing hardware and software facilities, realizing open-looping coordinated control of active and reactive power for the first time. The control strategy introduced can noticeably reduce network losses, enhance system stability and improve frequency and voltage dynamic quality, thus finally realizing harmonization among the targets of security, quality and economical operation of power system.

It should be noted that the ICOC system of AGC and AVC is put forward with a strong engineering background, bearing reference significance for other regional power grids. Additionally, the concept and research approaches of ICOC are also applicable for other industrial automatic control systems, especially in complicated manufacturing systems and some process optimizing systems.

ACKNOWLEDGMENTS

This work was supported by China's National High Technology Research and Development Program (2011AA05A112), China's

National Nature Science Foundation under Grant 50607011, science and technology projects of State Grid Corporation of China, as well as by the research projects of China's Hubei Electric Power Corp. and Liaoning Electric Power Co., Ltd.

REFERENCES

1 Zhou, J. Study on the real-time optimal dispatch in deregulated environments. Ph.D. Thesis, Tsinghua University, Beijing, China, 2005.

2 Wu, Q.; Popovic, D.H.; Hill, D.J. Voltage security enhancement via coordinated control. *IEEE Trans. Power Appar. Syst.***2001**, *16*, 127–135.

3 Corsi, S.; Pozzi, M.; Sabelli, C.; Serrani, A. The coordinated automatic voltage control of the Italian transmission grid—Part I: Reasons of the choice and overview of the consolidated hierarchical system. *IEEE Trans. Power Appar. Syst.***2004**, *19*, 1723–1732.

4 Kundur, P. *Power System Stability and Control*; Balu, N.J., Lauby, M.G., Eds.; McGraw-Hill: New York, NY, USA, 1994.

5 Ye, X.; Lu, Z.X.; Qiao, Y.; Li, J.; Wang, F.; Luo, W. A primary analysis on mechanism of large scale cascading trip-off of wind turbine generators. *Autom. Electr. Power Syst.* **2012**, *36*, 11–17.

6 Roncero, J.R. Integration is Key to Smart Grid Management. In Proceedings of Smart Grids for Distribution Conference, CIRED Seminar, Frankfurt, Germany, 23 June 2008; pp. 1–4.

7 Yu, X.Y.; Yu, J.L. A Dynamic power flow algorithm with joint regulation of active and reactive power. *Power Syst. Technol.* **2005**, *29*, 61–65.

8 Sun, W.Q.; Wang, C.M.; Zhang, Y.; Yu, G.Q.; Zhu, D.K. Active/reactive power coordinated optimization in power systems comprehensive energy saving. *Electr. Mach. Control* **2010**, *14*, 41–47.

9 Hu, W.; Wang, S.Y.; Xu, F.; Zhang, W.; Zhou, J. An optimal coordinated control strategy of AGC and AVC based on hierarchical control. *Autom. Electr. Power Syst.* **2011**, *35*, 40–45.

10 Yang, F.; Sun, Y.Z.; Cheng, L. Coordinated control of AGC redispatching and AVC to enhance voltage security.*Autom. Electr. Power Syst.* **2010**, *34*, 7–11.

11 Hu, W. Studies on hybrid automatic voltage control system. Ph.D. Thesis, Tsinghua University, Beijing, China, 2002.

12 Liu, W. *Power System Frequency Regulation and Automatic Generation Control*; China Electrical Press: Beijing, China, 2006.

13 Wright, M.H. The interior-point revolution in constrained optimization. In Proceedings of the First Pacific Conference on Mathematics, Hong Kong, China, January 1998; pp. 19–23.

14 Zhou, S.X.; Zhu, L.Z.; Guo, X.J.; Wang, X.H. *Power System Voltage Stability and Control*; China Electrical Press: Beijing, China, 2004.

15 Lefebver, H.; Fragnier, D.; Boussion, J.Y.; Mallet, P.; Bulot, M. Secondary coordinated voltage control system: feedback of EDF. In Proceedings of IEEE PES 2000 Summer Meeting, Seattle, WA, USA; 2000; pp. 290–295.

16 Guo, Q.L.; Sun, H.B.; Zhang, B.M.; Wu, W.C.; Li, Q. Study on coordinated secondary voltage control. *Autom. Electr. Power Syst.* **2005**, *29*, 19–24.

17 Conejo, A.; Aguilar, M.J. Secondary Voltage Control: Nonlinear Selection of Pilot Buses, Design of an Optimal Control Law, and Simulation Results. *IEE Proc. Gener. Trans. Distribut.* **1998**, *145*, 77–81.

CHAPTER 2

Using Exergy to Understand and Improve the Efficiency of Electrical Power Technologies

Marc A. Rosen[1,*] and Cornelia Aida Bulucea [2]

[1] Faculty of Engineering and Applied Science, University of Ontario Institute of Technology, 2000 Simcoe Street North, Oshawa, Ontario, L1H 7K4, Canada

[2] Faculty of Electromechanical and Environmental Engineering, University of Craiova, Romania

ABSTRACT

The benefits are demonstrated of using exergy to understand the efficiencies of electrical power technologies and to assist improvements. Although exergy applications in power systems and electrical technology are uncommon, exergy nevertheless identifies clearly potential reductions in thermodynamic losses and efficiency improvements. Various devices are considered, ranging from simple electrical devices to generation systems for electrical power and for multiple products including electricity, and on to electrically driven. The insights provided by exergy are shown to be more useful than those provided by energy, which are sometimes misleading. Exergy is concluded to have a significant role in assessing and improving the efficiencies of electrical power technologies and systems, and provides a useful tool for engineers and scientists as well as decision and policy makers.

KEYWORDS

exergy; electricity; power generation; energy conversion; electricity use

1. INTRODUCTION

The use of energy as a measure for understanding and improving the efficiencies of energy systems can be misleading and confusing. The quantity exergy can be used to assess and improve the efficiencies energy systems, and can help better understand the losses in energy systems by providing more useful and meaningful information than energy provides. Exergy is the maximum work (or electricity) producible from a system or a flow of matter or energy relative to a reference environment. Exergy is a measure of the potential of the usefulness or value of a system or flow. Exergy is also a measure of the potential to cause change, as a consequence of not being in equilibrium relative to the reference environment.

Exergy can clearly identify efficiency improvements and reductions in thermodynamic losses. Exergy can also identify better than energy the environmental benefits and economics of energy technologies. Thus, exergy has an important role to play in increasing efficiencies of energy systems and technologies.

Many feel that exergy is applicable only to systems or studies involving extensive thermodynamics, in areas like mechanical and chemical engineering. Thus the full value of exergy is not achieved because it is neglected or underutilized in other fields. One such area is electrical engineering, where exergy applications are uncommon.

One reason that exergy is not used extensively in electrical engineering studies is that general thermodynamic assessments often deal with many energy forms, e.g., electricity, work, heat, chemical, kinetic, potential, etc., while electrical engineering assessments often focus on electricity, and sometimes work. In such situations, energy and exergy flows are identical (since exergy is the work or electrical equivalent of energy), so efficiencies tend to be the same and the benefits of using exergy methods are less obvious. Nevertheless, there are benefits. First, losses are identified in terms of cause and located more accurately with exergy. Second, in systems where non-electrical quantities are involved, efficiencies on energy and exergy bases differ, sometimes markedly.

This paper goes on to describe exergy and exergy methods and their applicability to electrical systems. Examples are used to illustrate the use of exergy as a tool to understand and improve efficiency. The objective is to improve how efficiencies and losses are understood for electrical systems, thereby assisting efforts to improve them.

2. EXERGY AND EXERGY ANALYSIS

Exergy is defined as the maximum amount of work which can be produced by a system or a flow of matter or energy as it comes to equilibrium with a reference environment. Exergy is a measure of the potential of the system or flow to cause change, as a consequence of not being completely in stable equilibrium relative to the reference environment. Unlike energy, exergy is not subject to a conservation law (except for ideal, or reversible, processes). Rather exergy is consumed or destroyed, due to irreversibilities in any real process. The exergy consumption during a process is proportional to the entropy created due to irreversibilities associated with the process.

Exergy analysis [1,2,3,4,5,6,7,8,9] is a methodology that uses the conservation of energy principle (embodied in the first law of thermodynamics) together with non-conservation of entropy principle (embodied in the second law) for the analysis, design and improvement of energy and other systems. The exergy method is useful for improving the efficiency energy-resource use, for it quantifies the locations, types and magnitudes of wastes and losses. In general, more meaningful efficiencies are evaluated with exergy rather than energy analysis, since exergy efficiencies are always a measure of the approach to the ideal.

Therefore, exergy analysis identifies the margin available to design more efficient energy systems by reducing inefficiencies.

In exergy analysis, the characteristics of the reference environment must be specified completely. This is commonly done by specifying the temperature, pressure and chemical composition of the reference environment. The results of exergy analyses, consequently, are relative to the specified reference environment, which in most applications is modelled after the actual local environment. The exergy of a system is zero when it is in equilibrium with the reference environment. This tie between exergy and the environment leads to some of the implications regarding environmental impact that are discussed subsequently.

Exergy also can be applied beyond assessing efficiencies and losses, in areas such as economics [10,11,12] and environmental impact assessment [13,14,15]. The relationship between exergy and economics, particularly the trade-offs that normally occur between efficiency and costs, has been an important concern for decades and continues to be so. Economic methods based on exergy, e.g., exergoeconomics, have evolved and are applied in some areas. The environmental impacts and non-sustainability of energy use have become of significant concerns recently, and exergy methods can assist improvement efforts by (1) reducing environmental emissions and lengthening the lives of resources through increasing efficiencies, and (2) assessing the potential impacts of emissions. The linkages between exergy and economics and environmental impact are beyond the scope of this article and thus not considered further here.

Many engineers and scientists suggest that energy systems are best evaluated using exergy analysis because it provides more insights, especially for efficiency improvement, than energy analysis. Exergy analysis and its application to many processes and systems are discussed further elsewhere [1,2,3,4,5,6,7,8,9]. As energy policies increasingly play an important role in addressing development and sustainability issues, decision and policy makers would also benefit from an appreciation the exergy concept and its applications.

3. APPLICATIONS OF EXERGY TO ELECTRICAL SYSTEMS

Exergy can be applied to electrical systems in much the same way as it is to general thermodynamic systems. Usually, since there are fewer energy forms involved, the analyses are less involved and complex for electrical systems.

The range of devices that can be considered is extensive, ranging from generators and transformers to power plants. Here, we consider four categories of electrical systems:

- Simple electrical conversion devices.
- Electrical generation systems.
- Generation systems for multiple products including electricity.
- Electrical storage systems.
- Devices and systems that use or are driven by electricity.

Many examples can be used to demonstrate how exergy methods improve understanding of efficiencies and help increase them. Several examples in each of the above categories are presented in subsequent sections for illustrative purposes.

4. ILLUSTRATIONS FOR VARIOUS ELECTRICAL SYSTEMS

4.1. Simple Electrical Conversion Devices

Simple electrical conversion devices include transformers, alternators, generators, motors and static converters, and these are considered here. In such devices, electrical energy is being converted to other type of electrical energy, or work and electricity quantities are being interchanged.

Since electricity and work are the same on energy and exergy bases, the overall energy and exergy efficiencies for such devices are the same. The losses differ greatly on energy and exergy bases, however, as all energy losses are typically associated with waste heat. Most exergy losses are associated with internal consumptions of exergy due to irreversibilities and little with waste thermal exergy emissions.

Electrical losses often involve resistive losses, which appear as waste heat. Electric generator losses usually include losses in the winding and magnetizing losses in the core, plus rotational losses. Power transformer losses are usually of the same kind as electric generator losses. Transformer losses have two main components: (1) Core loss (or no-load loss), due to magnetizing and de-magnetizing the core during normal operation; this loss occurs whenever the transformer is energized and does not vary with load. (2) Coil loss (or load loss), which occurs in the primary and secondary coils of the transformer. This loss is a function of the resistance of the winding materials and which varies with the transformer load. Losses in these devices are also caused friction.

4.2. Electrical Generation

Idealized and actual electrical power plants are considered here, including thermal power plants, hydroelectric plants and fuel cells.

Ideal Thermal Power Plant

A Carnot heat engine represents an ideal power plant. Energy efficiencies for this device are counter-intuitive. For a Carnot power plant operating between a heat source at a temperature of 600 K and a heat sink at 300 K, for example, the energy efficiency is 50% (i.e., 1 – 300/600 = 0.5). Yet a Carnot engine is ideal. Clearly, the energy efficiency is misleading as it indicates that a significant margin for improvement exists when in fact there is none. Although a significant amount of waste heat is emitted, it is rejected at the temperature of the reference environment, where heat is freely available in abundance. Thus the waste heat is useless for processes. The exergy efficiency of an ideal device is 100%, properly indicating its ideal nature in a straightforward and clear manner. Zero exergy is associated with the waste heat, reflecting its value or usefulness.

Coal and Nuclear Power Plants

Actual coal-fired and nuclear electricity generation plants are considered.

First, we consider the coal-fired Nanticoke generating station, which has been operating since 1981 in in community of Nanticoke, Ontario (Canada). Each of the eight station units has net outputs of 505 MW of electricity. A single station unit (see Figure 1) can be divided into four main sections [15,16]:

- *Steam Generators*: Natural circulation steam generators combust pulverized coal to produce primary and reheat steam, and flue gas exits via chimneys.
- *Turbine Generators and Transformers*: Electricity is generated as steam passes through a turbine generator connected to a transformer. Each turbine generator has one single-flow high-pressure cylinder, one double-flow intermediate-pressure cylinder and two double-flow low-pressure cylinders. Steam exhausted from the high-pressure cylinder is reheated in the steam generator.
- *Condensers*: Steam exhausted from the low-pressure turbines enters the condenser, where cooling water condenses the steam.
- *Preheating Heat Exchangers and Pumps*: The pressure of the condensed steam is increased in a series of electrical pumps, while the temperature of the condensate is increased in a series of heat exchangers, with heat supplied from steam extractions from the turbines.

Overall balances of exergy and energy for the station are illustrated in Figure 2. The main findings, which improve understanding of the thermodynamic behaviour of the plant and help identify areas having significant efficiency-improvement potential, follow:

- The overall energy efficiency (ratio of net electrical energy output to coal energy input), is 37%, and the corresponding exergy efficiency 36%.
- The steam generators appear significantly more efficient on an energy basis (95%) than on an exergy basis (50%). Physically, this discrepancy implies that, although most of the input energy is transferred to preheated water, the energy is degraded as it is

transferred. Most of the exergy losses in the steam generators are associated with internal consumptions, mainly due to combustion and heat transfer.

- Large quantities of energy enter the condensers (about 775 MW per unit), of which close to 100% is rejected. A small quantity of exergy enters (about 54 MW per unit), of which about 25% is rejected and 75% internally consumed.
- Energy losses in other plant devices are very small (about 10 MW). The exergy losses are moderately small (about 150 MW), and are mainly associated with internal consumptions.

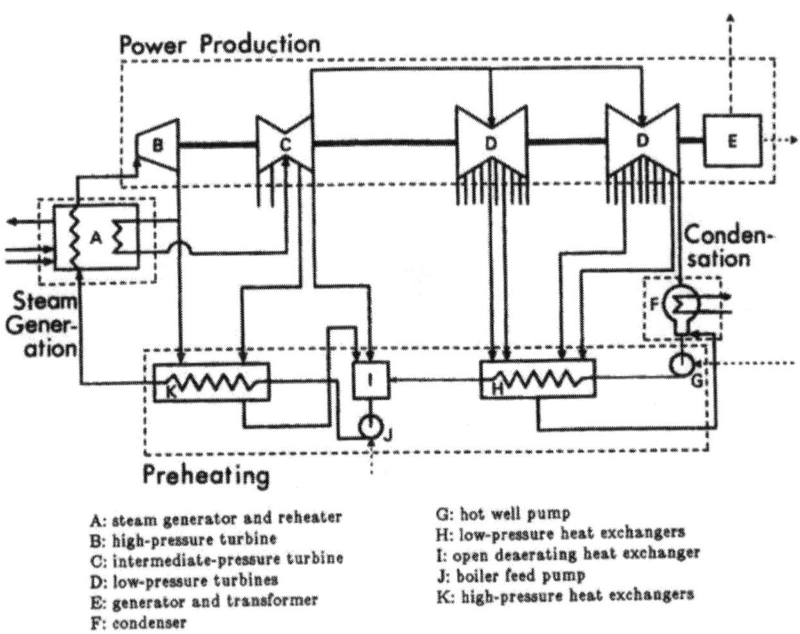

Figure 1. Breakdown of the electrical generating station unit considered into four main sections. The external inputs are coal and air, and the output is stack gas and solid waste for unit A. The external outputs for unit E are electricity and waste heat. Electricity is input to units G and J, and cooling water enters and exits unit F.

Similar results are observed for a nuclear power plant, the main difference being that the large exergy consumptions occur in a nuclear reactor instead of a coal combustor. In the nuclear plant, the main losses are associated with the fission reactor, the turbines, the heat exchangers, the pumps and the electrical generator and transformer.

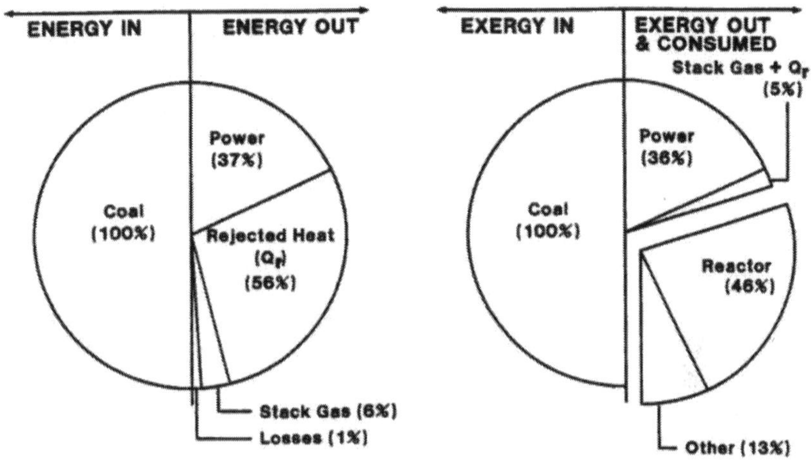

Figure 2. Overall energy and exergy balances for the station. Energy balance inputs and outputs of energy (left). Exergy balance showing inputs and outputs and consumptions of exergy (right).

Hydroelectric Plant

For a hydroelectric generating station, the potential energy of elevated water is converted into electricity. As potential energy and potential exergy are identical, the overall energy and exergy efficiencies are the same. However, the energy losses exit mainly with the water leaving the system, while the main exergy losses are associated with internal consumptions. In these analyses, the reference environment elevation is taken to be that at the bottom of the system.

Fuel Cell

For a fuel cell system, electricity is generated electrochemically as the inputs, hydrogen and oxygen, react to form water. As the energy and exergy values of hydrogen differ somewhat, the energy and exergy efficiencies differ slightly for these systems. But the energy and exergy losses differ, the main exergy losses being due to internal consumptions and the main exergy losses associated with waste heat. The main losses are associated with the fuel cell stack, the expanders, the heat exchangers and the pumps.

Wind Turbine System

For a wind generating station, the kinetic energy of moving air is converted to electricity. As potential energy and kinetic exergy are identical, the overall energy and exergy efficiencies are the same. However, the energy losses exit mainly with the air leaving the device, while the main exergy losses are associated with internal consumptions. In these analyses, the reference environment for velocity is taken to be a non-moving point on the earth.

Solar Photovoltaic System

For a solar photovoltaic generating station, the energy of sunlight is converted directly to electricity. As the energy and exergy of sunlight are very close, they are assumed identical here. Thus, the overall energy and exergy efficiencies are the same. However, the energy losses exit mainly with the waste heat leaving the system, while the main exergy losses are associated with internal consumptions.

Solar Thermal Power Generation

For a solar thermal power generation, the energy of sunlight is converted indirectly to electricity, using a thermal power cycle that often is similar to a Rankine cycle. Again, the overall energy and exergy efficiencies are the same since the energy and exergy of sunlight are assumed identical, and energy losses exit mainly with the waste heat emissions while internal consumptions are responsible for the main exergy losses.

4.3. Cogeneration and Trigeneration

Cogeneration and trigeneration systems are considered. These systems generate multiple products including electricity.

Cogeneration Systems

Cogeneration, or combined heat and power (CHP), systems generate co-products: electricity and useful heat. The energy and exergy efficiencies for such systems vary due to the thermal product, for which energy and exergy differ. The energy and exergy efficiencies are nearer when the temperature of the product heat is higher, and deviate more as the temperature of the product heat decreases toward the reference environment temperature. The energy and exergy losses for such systems are identified to be in different locations and due to different causes. The main losses are associated with the furnace and boiler, the turbines, the heat exchangers, the electrical generator and transformer, the stack gas and the pumps.

Trigeneration Systems

Trigeneration systems are extended cogeneration systems which generate a cold product in addition to electricity and useful heat. The energy and exergy efficiencies for such systems vary due to the hot and cold thermal products. For each of these commodities, energy and exergy differ. The energy and exergy efficiencies are deviate more when the temperatures of the product thermal quantities are further removed from the reference-environment temperature (i.e., the hot product is hotter and the cold product colder). The energy and exergy efficiencies are nearer as the temperatures of the thermal quantities approach the reference-environment temperature. There are particular difficulties in evaluating the energy efficiencies of such systems due to the cold product, and the difficulty in assessing its energy. The exergy of cold are hot thermal products are straightforwardly evaluated, so meaningful exergy efficiencies are readily determined.

As for cogeneration systems, the energy and exergy losses for such systems are identified to be in different locations and due to different causes. The main losses are associated with the furnace and steam generator, the turbines, the heat exchangers, the stack gas, the electrical generator and transformer, and the chiller.

4.4. Electrical Storage Systems

Electrical storage systems are considered here, the most common of which are batteries. The energy efficiency on a fractional basis of an electrical storage device is usually determined as the ratio of the quantity of electricity obtained from the storage while discharging divided by the electrical energy supplied to the storage while charging. For a battery, the energy loss is observed as heat, which warms the battery and the surroundings. Low charging and discharging rates help keep a battery cool and improve the battery life. Given that the inputs and outputs are both electricity for electrical storage systems, the energy and exergy efficiencies are equal.

4.5. Electrically Driven Devices and Systems

Space and Hot Water Heaters

Space and hot water heating can be accomplished in many ways. Two competing technologies are considered here:

- *Electrical Resistance Space Heater*: In an electrical resistance space heater, almost all of the electricity that enters the unit is dissipated to heat within the space. Thus the energy efficiency is nearly 100% and there are almost no energy losses. Yet the exergy efficiency of such a device is typically less than 10%, indicating that the same space heating can in theory be achieved using one-tenth of the electricity.

- *Heat Pump*: Some of the savings in electricity use identified in the previous point can be attained using a heat pump. The use of even a relatively inefficient heat pump can reduce the electricity used to achieve the same space heating by one-third. A high-efficiency heat pump can achieve even greater reductions electricity usage. The main losses occur in the motor and compressor, the condenser, the heat exchangers and the throttle valve.

Efficiencies for hot water heating are also markedly different for different technologies and losses differ in characteristics and amount. For electrical hot water heaters, the efficiencies are low and wastes are mainly attributable to waste heat (on an energy basis) and to internal consumptions (on an exergy basis). For natural gas-fired, as an alternative, the results are similar, although the technologies differ.

Clearly the use of energy efficiencies and losses is quite misleading for heating.

Industrial Furnaces and Heaters

Energy efficiencies for industrial process heating vary slightly depending on the heat retention characteristics of the heater. Exergy efficiencies for industrial process heating vary significantly, depending primarily on the temperature of the product heat. Typically, the exergy efficiency is high when a high-quality input such as electricity is used for a high-quality task (like producing high-temperature heat). Thus, the use of energy efficiencies and losses is quite misleading for industrial heating.

A breakdown of the exergy losses for such systems often reveals that the exergy losses are divided among those due to irreversibilities associated with the combustion reaction (even if done adiabatically) and those due to heat transfer from the peak theoretical combustion temperature to the actual temperature of the working fluid.

Refrigeration Systems

Refrigeration systems operate like heat pumps, but with the heat removal from the cold space, rather than the warmer heat rejected, being the product. As for trigeneration systems which involve a cold product, the energy and exergy efficiencies for such systems vary due to the dependence of the exergy of the cold thermal product on temperature. The energy of the cold product is also difficult to determine, so the energy-based figure of merit used for refrigeration systems is the coefficient of performance (COP), which is simply the energy efficiency divided by 100. The energy and exergy efficiencies differ markedly for refrigeration systems, but the exergy efficiencies are more intuitively meaningful.

The energy and exergy losses for such systems are identified to be in different locations and due to different causes. The main losses occur in the motor and compressor, the condenser, the heat exchangers and the throttle valve.

Pumps and Fans

Electric pumps and fans convert electricity to mechanical energy in the form of pressure rise and kinetic energy. The energy and exergy of the products are usually similar, leading to similar device efficiencies.

Lights

Not only does lighting consume a significant amount of energy in society, it is also extremely inefficient. Light bulbs convert electricity to visible light. As the product electromagnetic radiation can be taken to have similar energy and exergy contents, the energy and exergy efficiencies of lighting are similar. Lighting efficiencies differ markedly from the efficiencies of many home appliances. The potential for energy and exergy savings from lighting is large, due to its inefficiency and large contribution to societal energy use.

Home Electric Appliances for Heating-Based Services

Electric household appliances for heating services (e.g., ovens, toasters, clothes dryers) have characteristics of the heating devices discussed earlier in this section, with the main differences dependent on the temperature of the product heat. The exergy efficiencies also depend on the temperatures of the waste heat emissions.

Electric Transportation Systems

Electric railways trains supplied by AC or DC contact lines are equipped with three-phase induction motors and variable voltage and frequency inverters [26,27]. Using structural diagrams and high-performances converters, appropriate vehicle control can be achieved so the optimum traction and minimum electrical energy use are ensured. In such devices, the energy and exergy efficiencies are equal, as the product kinetic energy and kinetic exergy are equivalent. With the same induction machines and use of appropriate controls, these drive systems can realize both traction and electric braking as part of the operation of the electric traction vehicle.

5. NUMERICAL EXAMPLES

Typical values of energy and exergy efficiencies for the electrical devices and systems considered in the previous section are given in Table 1. In addition, exergy losses are broken down in that table into internal exergy consumptions and waste exergy emissions.

An examination of the efficiencies in Table 1 shows that in some instances the energy and exergy efficiencies are the same, while in other cases they differ significantly. Where they differ, the exergy efficiencies provide a measure of how nearly the process approaches ideality and the energy efficiencies can cause confusion. Where the efficiencies are similar, the energy efficiencies can be treated as good approximations of the more meaningful exergy efficiencies. The efficiencies tend to be similar where the energy input and product energy outputs are of the same quality. The efficiencies differ when energy inputs and outputs are of significantly different quality (e.g., electricity as opposed to heat at high temperature as opposed to cold thermal energy).

Of the electrical resistance heaters, the industrial has the higher exergy efficiency because it delivers heat of a higher quality, as reflected by its temperature. Space and hot water heaters deliver lower quality heat.

The refrigeration and heat pump systems are difficult to assess using energy efficiencies as the values exceed 100%. The energy efficiency for trigeneration is not a true efficiency as it includes the cold product energy, which is normally evaluated with a coefficient of performance (COP) as a figure of merit. In both these cases, exergy efficiencies are rational and meaningful.

The energy loss can be found using the corresponding energy efficiency as 100% minus the energy efficiency. The exergy loss can be found similarly. The overall energy loss is made up entirely of waste energy emissions from the system. The overall exergy loss differs significantly, as it consists of both waste exergy emissions from the system and internal exergy consumptions (or destructions) due to irreversibilities.

Table 1. Energy and exergy efficiencies and breakdown of exergy losses for selected electrical devices.

Device	Efficiency (%)		Breakdown of exergy loss (%)*	
	Energy	Exergy	Consumption	Emission
Simple electrical conversion devices				
Transformer (high power)	93–98	93–98	73	27
Transformer (low power)	75–90	75–90	73	27
Alternator (low power)	70–73	70–73	73	27
Generator (high power)	90–95	90–95	73	27
Motor	80–87	80–87	73	27
Static converter	97–99	97–99	73	27
Electrical generation systems				
Ideal thermal power plant	64	100	–	–
Coal–fired power plant	37	36	92	8
Nuclear power plant	30	30	98	2
Hydroelectric power plant	90	90	88	12
Fuel cell system	33	33	89	11
Wind turbine system	80–97	80–97	97	3
Solar photovoltaic system	6–25	6–25	88	12
Solar thermal power generation	10–30	10–30	70	30
Cogeneration and trigeneration systems				
Cogeneration system	74	31	99	1
Trigeneration system	94	28	92	8
Electrical storage systems				
Battery (lead–acid)	75–85	75–85	92	8
Battery (lead–acid) with charging circuit**	45–68	45–68	92	8
Electrically driven devices and systems				
Resistance space heater	99	6	97	3
Heat pump	380***	19	94	6
Hot water heater	90	10	92	8
Industrial heater	70	51	70	30
Refrigeration system	310***	24	95	5
Pump	70	70	97	3
Fan	90	90	97	3
Lighting (incandescent)	5	5	92	8
Lighting (fluorescent)	20	20	97	3
Electric oven (home)	70–85	50–55	92	8
Toaster	70	40	92	8
Clothes dryer	60	30	92	8
Electric transportation system (traction)	80	80	92	8
Urban electric transportation system (electric rheostatic braking)	80	30	70	30
Urban electric transportation system (electric braking with recovery)	80	73****	99	1

* Breakdown values indicate percentage of overall exergy loss for the device, and thus sum for each device to 100%.

** Assumes charging circuit has an energy efficiency between 60 and 80%.

*** Conventionally expressed as coefficient of performance (COP) by dividing by 100.

**** Actual electric transportation systems use static converters and induction machines to facilitate electric regenerative braking, with recovered energy provided to other operating trains and increasing energy and exergy efficiencies.

An examination of the breakdown of exergy losses in Table 1 shows that in most instances, the exergy losses are mainly associated with internal exergy consumptions. The waste exergy emissions tend to be smaller, usually less than 30% of the total exergy loss.

Table 1 is a compilation of data from different sources, which has been modified to be comparable and illustrative. Some details on the values presented in Table 1 are as follows:

- The reference-environment temperature is taken to be 20 °C.
- Energy efficiency values are presented for the transformer, generator, alternator motor and static converter which are typical of some types. Heat losses from the transformer, generator, alternator and motor are taken to occur at 130 °C. These values can also vary markedly among devices.
- Data for the coal-fired plant is based on the Nanticoke generating station [15,16].
- Data for the nuclear plant is based on the Pickering nuclear generating station [16], located in Pickering, Ontario (Canada). The input is taken to be the net heat of fission delivered to the reactor (rather than the uranium).
- The ideal thermal power plant is taken to be operating using a heat source at a temperature of 550 °C, which roughly corresponds to the temperature of the high-pressure superheated steam exiting the steam generator of the coal-fired Nanticoke generating station [15,16].
- For the hydroelectric power plant, water is assumed to exit at the elevation of the reference environment with negligible kinetic energy, thus having zero exergy and energy. It is also assumed that waste heat is released at an average temperature of 60 °C.
- The fuel cell system is based on an automotive fuel cell system assessed previously by Cownden *et al.* [18]. The system includes the fuel cell stack and the balance of plant, which includes pumps, compressors, heat exchangers, etc.
- Wind turbine system energy efficiencies vary significantly with wind speed, and are usually designed for highest efficiency operation at a

given wind speed. Values are reported here for typical operating energy efficiencies for wind turbine systems at wind speeds that are reasonable in terms of systems operation. The main component of the system is the wind turbine itself, which has energy efficiencies typically ranging from 75% to 94% [20,21]. The device is assumed to emit waste heat at 30 °C, a temperature near the ambient condition.

- Energy efficiencies values for photovoltaic electricity generation are based on confirmed efficiencies for terrestrial solar cells and modules, measured under the global AM1-5 spectrum (1,000 W/m^2) at 25 °C [22,23]. These devices are assumed to emit waste heat at 60 °C, a temperature between ambient conditions and waste heat flows for industrial heating.

- Solar thermal energy conversion devices use the energy in sunlight to heat a working fluid to a temperature sufficiently high to drive a thermal power cycle. Such systems usually utilize concentrating solar collectors so that reasonably high temperatures can be attained. The mean annual efficiencies reported are about 10%–20% for systems built over the past two decades and 30% for more recent systems [23]. The devices are taken to emit waste heat at a temperature of 145 °C, which is typical for industrial heating systems and some solar thermal systems.

- The cogeneration plant is based on the steam cycle-based plant analyzed in Section 6.3 of Kotas [2]. The plant has an electrical power output of 6 MW and a useful thermal output of 25 MW delivered at a mean temperature of 133 °C. The high-pressure steam temperature and pressure are 400 °C and 40 bar, respectively.

- The trigeneration system is based on a design for a cogeneration-based heating and cooling project in downtown Edmonton, Alberta (Canada) [17]. The system design has an initial supply capacity of 230 MW (thermal) for heating and 100 MW (thermal) for cooling and used electric centrifugal chillers. The efficiency data are based on a previous analysis by Rosen *et al.* [17]. The exergy loss breakdown is assumed the same as for the coal power generation example.

- The electrical storage systems considered are common lead-acid batteries, without and with a charging circuit. The energy efficiency represents the percentage of electricity supplied to the battery system that is recovered on discharging. Energy efficiency values for typical lead-acid batteries and charging circuit are obtained from a company [19]. For the battery, the charging circuit and the combination of the two, the exergy efficiency is identical to the energy efficiency. The heat loss from the battery is assumed released at 45 °C.

- The electrical resistance space heater is assumed to deliver product heat at 40 °C, and to emit waste heat at 30 °C.

- The electrical hot water heater is assumed to deliver product heat at a temperature of 55 °C, and to emit waste heat at 45 °C.

- The electrical industrial heater is assumed to deliver product heat at a temperature 800 °C, and to emit waste heat at 145 °C.

- The refrigeration system is based on the vapour-compression plant analyzed in Section 5.4 of Kotas [2]. The plant, which uses ammonia as the working fluid, has a refrigerating duty of 93 kW and maintains the cold space at –1 °C. The waste from the refrigeration system is taken to be the heat released from evaporator, which is assumed to occur at 35 °C. The plant efficiency is relatively low compared to systems available presently.

- The heat pump taken to be same as refrigeration system, except that the heat delivered is the product. Waste from the heat pump system includes miscellaneous heat releases from all devices and assumed to be small (10% of the exergy loss), which are assumed to occur at 40 °C. The plant efficiency is relatively low compared to available systems.

- Pump and fan efficiencies vary greatly, depending on capacity and other characteristics of the device. Values are taken for pumps used in variable air volume systems for an office building [25], while typical values are used for fans [24]. Waste heat emissions for both devices are assumed to occur at 30 °C.

- Lighting efficiencies vary significantly. Incandescent light bulbs and compact fluorescent lamps (a phosphor-coated gas discharge tube) convert about 5% and 20%, respectively, of the electricity they use to visible light [24]. Waste heat emissions from incandescent bulbs are assumed to occur at 45 °C and from compact fluorescent lamps at 30 °C.

- The home electric appliances for heating-based services considered here are electric ovens, clothes dryers and toasters. All convert electricity to useful heat with a typical energy efficiency of at least 60% [24]. The product heat and waste heat emission temperatures vary, depending on the device. The device temperatures are taken to be between the temperatures for industrial and hot water heating, while waste heat emissions are all assumed to occur 45 °C, for simplicity.
- The electric transportation systems on the basis of variable voltage and frequency static converters and induction machines facilitate improved electric braking, with energy recovery. In that operating regime, the vehicle provides energy to the DC network through an inverter. The recovered energy is used by other running trains, increasing energy and exergy efficiencies. For the traction-based electric transportation system, the energy and exergy efficiencies are calculated on basis of the efficiencies of static converters, electric transformers and induction motors. For the urban electric transportation system with electric braking, braking regimes with and without energy recovery are shown. The exergy efficiency is low for the system with rheostatic braking and high for the case of braking with energy recovery. The braking regime is only of significance for urban vehicles, when it is performed often.

6. DISCUSSION

The efficiencies and losses for the electrical technologies and systems considered in this article can be compared. Clearly the energy efficiencies of such systems are inconsistent, in that some use a simple efficiency, others a coefficient of performance and others a different figure of merit. Also, the energy losses are misleading, in that they focus attention on areas where the main deviations from ideality are not occurring.

The exergy efficiencies and losses describe the efficiencies of the electrical technologies and systems better and provide more meaningful information. With these insights, a better understanding is attained of the factors affecting efficiency, and efforts to improve performance can be better allocated and directed.

In particular, low values of exergy efficiencies, or high exergy losses, indicate that a significant margin for efficiency improvement exists in theory. Realizing this efficiency improvement requires ingenuity and creativity, and involves trade offs with other factors such as economics and environmental impact. The breakdown in exergy losses allows the causes, locations and magnitudes of the exergy losses to be understood better, so that efficiency-improvement efforts can focus more directly on those factors likely to be causing efficiency losses.

7. CONCLUSIONS

The benefits of using exergy to understand the efficiencies of electrical technologies systems and to guide improvement efforts have been demonstrated. It is concluded that the concepts encompassing exergy have a significant role to play in evaluating and increasing the efficiencies of such technologies. Exergy should prove useful in such activities to engineers and scientists, as well as decision and policy makers.

ACKNOWLEDGEMENTS

Financial support for this research was provided by the Natural Sciences and Engineering Research Council of Canada.

REFERENCES AND NOTES

1 Moran, M.J. *Availability Analysis: A Guide to Efficient Energy Use*, revised ed.; American Society of Mechanical Engineers: New York, NY, USA, 1989.

2 Kotas, T.J. *The Exergy Method of Thermal Plant Analysis*, reprint ed.; Krieger: Malabar, FL, USA, 1995.

3 Moran, M.J.; Sciubba, E. Exergy analysis: principles and practice. *J. Eng. Gas Turb. Power* **1994**, *116*, 285–290.

4 Szargut, J.; Morris, D.R.; Steward, F.R. *Exergy Analysis of Thermal, Chemical and Metallurgical Processes*; Hemisphere: New York, NY, USA, 1988.

5 Szargut, J. International progress in second law analysis. *Energy* **1980**, *5*, 709–718.

6 Edgerton, R.H. *Available Energy and Environmental Economics*; D.C. Heath: Toronto, Canada, 1992.

7 Dincer, I.; Rosen, M.A. *Exergy: Energy, Environment and Sustainable Development*; Elsevier: Oxford, UK, 2007.

8 Sato, N. *Chemical Energy and Exergy: An Introduction to Chemical Thermodynamics for Engineers*; Elsevier: Oxford, UK, 2005.

9 Brodyanski, V.M.; Sorin, M.V.; Le Goff, P. *The Efficiency of Industrial Processes: Exergy Analysis and Optimization*; Elsevier: London, UK, 1994.

10 Yantovskii, E.I. *Energy and Exergy Currents: An Introduction to Exergonomics*; Nova Science Publishers: New York, NY, USA, 1994.

11 Mazur, V.A. Fuzzy thermoeconomic optimization. *Int. J. Exergy* **2005**, *2*, 1–13.

12 Dincer, I.; Rosen, M.A. Thermoeconomic analysis of power plants: An application to a coal-fired electrical generating station. *Energ. Conver. Manage.* **2003**, *44*, 2743–2761.

13 Connelly, L.; Koshland, C.P. Two aspects of consumption: Using an exergy-based measure of degradation to advance the theory and implementation of industrial ecology. *Resour. Conserv. Recycl.* **1997**, *19*, 199–217.

14 Szargut, J. *Exergy Method: Technical and Ecological Applications*; WIT Press: Southampton, UK, 2005.

15 Dincer, I.; Rosen, M.A. Exergy analysis of waste emissions. *Int. J Energ. Res.* **1999**, *23*, 1153–1163.

16 Rosen, M.A. Energy- and exergy-based comparison of coal-fired and nuclear steam power plants. *Exergy* **2001**, *1*, 180–192.

17 Rosen, M.A.; Le, M.N.; Dincer, I. Efficiency analysis of a cogeneration and district energy system. *Appl. Therm. Eng.***2005**, *25*, 147–159.

18 Cownden, R.; Nahon, M.; Rosen, M.A. Exergy analysis of a fuel cell power system for transportation applications.*Exergy* **2001**, *1*, 112–121.

19 *Sealed Lead Acid Battery Application*; Transtronics, Inc.: Lawrence, KA, USA. Available online: http://xtronics.com/reference/ batterap.htm. (Accessed on 24 June 2009).

20 Hau, E. *Wind Turbines: Fundamentals, Technologies, Application, Economics*, 2nd ed.; Springer: New York, NY, USA, 2006.

21 Polinder, H.; van der Pijl, F.F.A.; de Vilder, G.-J.; Tavner, P.J. Comparison of direct-drive and geared generator concepts for wind turbines. *IEEE Trans. Energy Convers.* **2006**, *21*, 725–733.

22 Green, M.A.; Emery, K.; Hisikawa, Y.; Warta, W. Solar cell efficiency tables (version 30). *Progress in Photovoltaics: Research and Applications* **2007**, *15*, 425–430.

23 Crabtree, G.W.; Lewis, N.S. Solar energy conversion. *Phys. Today* **2007**, *60*, 37–42.

24 Humphreys, C.J. Solid-state lighting. *MRS Bull.* **2008**, *33*, 459–470.

25 Wei, Z.; Zmeureanu, R. Exergy analysis of variable air volume systems for an office building. *Energ. Conv. Manage.***2009**, *50*, 387–392.

26 Mohan, N.; Undeland, T.M.; Robbins, W.P. *Power Electronics, Converters, Applications and Design*, 2nd ed.; John Wiley & Sons: New York, NY, USA, 1997.

27 Perticaroli, F. *Sistemi Elettrici per i Transporti: Trazione Elettrica (Electrical Systems for Transportation: Electric Traction)*; MASSON Publishing House: Barcelona, Spain, 1994.

CHAPTER 3

Analysis of Distributed Generation Systems, Smart Grid Technologies and Future Motivators Influencing Change in the Electricity Sector

Nur Asyik Hidayatullah, Blagojce Stojcevski, Akhtar Kalam

School of Engineering and Science, Victoria University, Melbourne, Australia.

ABSTRACT

The global Electricity Sector and its customers are faced with a number of challenges that are unparalleled since the advent of widespread electrification. Challenges including climate change, escalating energy prices, energy security and energy efficiency are converging to drive fundamental change in the way energy is produced, delivered and utilized. The electricity system of the future must produce and distribute electricity that is reliable, affordable and clean. To accomplish these goals, both the electricity grid and the existing regulatory system must be smarter. This paper explores smart grid technologies, distributed generation systems, R&D efforts across Europe and the United States, and technical, economical and regulatory barriers facing modern utilities.

KEYWORDS

Climate Change, Distributed Generation, Electricity Sector, Smart Grid

1. INTRODUCTION

The Electricity Sector for the greater part of the 20th century, and yet into the beginning of the 21st century, has relied upon the mainstream of its electricity being produced by strongly built centralized power stations that are connected to an abundance of unified transmission networks [1]. Regardless, crucial developments during the course of the next 10 - 15 years in terms of the innovation of the Electricity Sector from a centralized to decentralized power systems, architecture is a major priority for all governments and industry leaders in both developed and developing countries alike. This impetus has gained overwhelming momentum through constant bombardment of media personnel on intricate issues related to climate change, escalating energy prices, energy security and energy efficiency.

Moreover, changes in the Electricity Sector must also concentrate towards capitalizing on intelligent electronic infrastructures, international standards and decentralized substation automation initiatives in order to reach a robust and capable smart grid future. Smart grid technologies are capable of rapidly detecting, analyzing and responding to various perturbations in the network by integrating advanced control methods and Flexible AC Transmission Systems (FACTS). This paper aims to provide a state-of-the-art analysis of the principal issues that influence future change in the Electricity Sector; engineering challenges associated with the coordination and integration of smart grid technologies; and an illustrative view of distributed generation.

2. PRINCIPLE FACTORS INFLUENCING CHANGE IN THE ELECTRICITY SECTOR

This section of the paper intends to highlight some of the key issues influencing the Electricity Sector both at present and in the foreseeable future. Although many of these issues are interconnected in some form or other, they are intentionally discussed separately in order to establish a better insight into the influences each individual concern has on the electricity market.

2.1. Climate Change

The threat of climate change both in Australia and abroad is the main contributing factor prompting change in the Electricity Sector. The Intergovernmental Panel on Climate Change (IPCC) classifies climate change as the variation in the global atmospheric properties, either directly or indirectly caused by anthropogenic influences [2]. Systematic efforts executed through statistical analysis, R&D efforts and financial support is essential to establish new and innovative methods to reduce greenhouse gas emissions. If such mitigation actions are not taken seriously, then price wars instigated by the scarcity of fossil fuels, oil and gas resources will become apparent [2].

Problematic still is the economic limit imposed by the accessibility of fossil fuels and their extraction from diminishing locations. Plenty of debate on whether or not the delay of fossil fuel supplies will lead to reduced greenhouse gas emissions in the near future is a hot topic given that supplies are believed to run out in 35 - 40 years [3]. Consequently, dependable mitigation efforts must be contemplated to offer the world economy prospects in better improving the economic limit of fossil fuel accessibility, without increasing the goods and service standard of society.

However, climate change mitigation is still opposed by many factions around the world due to risks and uncertainties connected to the mitigation issue. When climate change initially emerged as a global phenomenon there was relentless cynicism that the scientific evidence provided by academics and scholars was inconsistent [4]. These outlooks were principally based on the assumption that most of the proxy data indicators analyzed by climatologists, geologists and paleontologists over the past 300 or so years could not be relied upon [4].

In spite of this, continuous monitoring of weather patterns and visible confirmation of worldwide disasters has opposed much skepticism on the issue. Events such as bushfires, coral bleaching, drought, floods, ice sheet melt, increased cyclone activity, ocean rises, species extinction and temperature variations all give substance to climate change [2]. **Figure 1**, illustrates fluctuating temperature variations of the Earth since 1900. The graph clearly indicates that from about 1970, the air temperature intensified to much hotter conditions by about 0.76°C ± 0.19°C [2].

Australia above all countries has committed itself to a pragmatic emissions target between 450 and 550 ppm CO_2-e [2]. International frameworks such as the Bali Roadmap, Kyoto Protocol and Copenhagen treaty play major roles within such trajectories and offer emission permits to nations that plummet beneath their targeted goals. These emission permits have quite a few benefits including lower abatement costs, unbiased trading coverage, enhanced financial security and improved profit enticements from trade earnings [2]. Conversely, they too have downfalls such as the loss of direction in mitigation policy attained from moving into international markets [2].

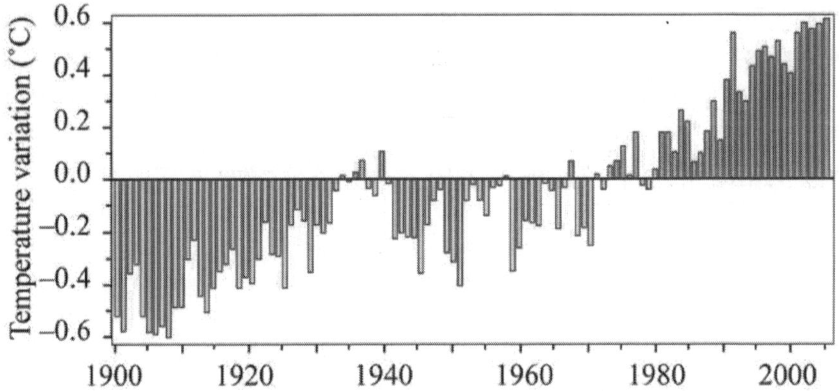

Figure 1. Annual temperature variation plot.

If no mitigation actions are taken to prevent the future state of climate change then society as we know it will definitely face grim circumstances. Emissions from toxic gases such as carbon dioxide (CH_2), methane (CH_4) and nitrous oxide (NO_4) will continue to pollute the atmosphere at dangerously high levels. Statistical evidence already proves this point by acknowledging the fact that carbon dioxide has exceeded the natural range over the last two million years by 25%, methane by 120% and nitrous oxide by 9% [2].

2.2. Escalating Energy Prices

Energy prices since the turn of the millennium have increased drastically as a direct result of higher fossil fuel demand and constraints in production growth. Resources such as crude oil and natural gas have significantly soared in price, but coal in particular has skyrocketed in value due to restrictions in mining and transportation [2]. **Figure 2**, illustrates a graphical representation of the price discrepancies of the three mentioned resources (oil, natural gas and coal) from the period of 1986 to 2011. As evident from the plot, each resource considerably intensifies in cost from about the year 2000 onwards.

If energy prices continue to follow this suit, then countless number of fossil fuel consumers will be in search of new and innovative substitutes which can adequately fulfill their energy generation needs at lesser costs. These replacements will ultimately reduce carbon dioxide emissions, but will also genuinely tackle the issue of climate change. One possible substitute that may be used is that of hydrocarbons, which expand on the concept of mimicking artificial liquid and coal products [2].

However, for the time being it is expected that in the up-and-coming years coal prices will plummet to record lows due to noticeable links between reduced product growth and return on investment. Renewable energy technology initiatives will then become more reasonably priced and affordable. This comes with the opinion that high-carbon fuel technologies will continue to proliferate on account of elevated oil prices [5]. As for energy prices in general, they are believed to differentiate even more so into the future among people living in inner and outer suburbs [2].

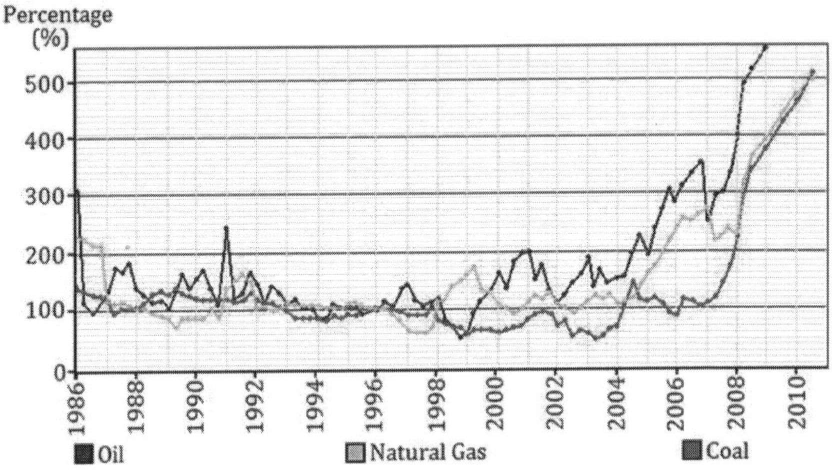

Figure 2. Energy price discrepancy plot.

In addition, the impacts of low-income households should also be considered. In most circumstances, poorer earning families will be reluctant to invest in more energy efficient technologies as they barely have enough income to support themselves. Even though they may be aware of the benefits such eco-friendly technologies provide (especially in terms of reduced energy bills), the option is just not viable for such individuals. Most of these families anyhow live in rental properties, thus implying they have no intention to install such energy saving luxuries unless the landlord decides to provide them out of their own pocket.

2.3. Energy Security

Ever since the aftermath of the September 11 attacks, energy security as a whole has become a major priority for the Electricity Sector. This comes amid concerns that a similar act of violence may well result in devastating effects within areas such as communication, electricity, gas or transportation [6]. This outcome considerably will slow down the global economy as a whole, but moreover impinge on the everyday processes of big business and society.

However, if such terrorization were to transpire on the Electricity Sector at this present time, it will not be through physical disruption or detonation of plant equipment. Instead the most likely cause of violation in terms of energy security will come from cyber threats that are brought upon by the encryption of communication systems within plant devices. This is to a great extent attributable to the vision that all power station equipment is coupled with some type of Intelligent Electronic Device (IED), SCADA system or IT infrastructure [6].

For this very reason, it is therefore imperative that power companies within the Electricity Sector invest both in time and resources by employing specialist engineers to continuously upgrade standards and procedures that may support the prevention of cyber security threats [6]. Companies should also ensure that fundamental precautions are fitted for the screening and education of staff members, but also make sure that auxiliary equipment is purchased in case of an emergency [6].

Similarly, wireless systems too face a risk of cyber intrusions. A large number of electricity companies make use of wireless systems in their networks as they believe features such as firewalls will protect their plant devices from prohibited users [6]. However, what many companies fail to recognize is that without the installation of extra security applications, the most amateur of hackers can access a company's personal data in a number of minutes via the use of a laptop computer and downloadable software tools from the internet [6].

This reality for the most part is based on the notion that IED and SCADA systems when initially designed never took onboard the concept of security. At first engineers believed that such systems will be used alone and not with other devices within the network; and that only sanctioned personal will have the right of entry to valuable system data [6]. However, this was not so the case. Currently, SCADA systems transmit data by means of modems. These modems behave as an additional pathway for hackers to access confidential data given that each modem dials out a specific dialog, which if harnessed correctly can scrutinize system logins.

Even though modern encryption topologies can be used alongside SCADA systems to prevent hackers from entering unprotected modems, the difficulty of implementing such topologies within the existing centralized power systems architecture lies on the impression that newly developed standards must be formulated first and foremost. The development of these new standards will comprise of dedicated encryption devices inbuilt to the SCADA remote terminal units (RTUs), but also to the modems [6]. If a hacker then attempts to enter the network through these modems, any data being transmitted at that particular instant of time will be publicized in the form of an encrypted message.

Adhering to this pragmatism, the development of newly planned standards will without doubt bring about major setbacks to the Electricity Sector. Dilemmas associated with the implementation costs of these standards will be a major concern, seeing as funds from company shareholders are essential for future growth. A further setback is the development of a transition period where companies not presently using new standards are given the opportunity to transform their operations within a fixed time slot. In this means companies are allowed to continue their normal operations without punishment, until the fixed time slot is surpassed.

Company mergers too can result in setbacks within the Electricity Sector. For instance, if two companies were to merge into the one organization, they will have difficulties interfacing their previously used cyber security protocols into the new system. This will lead to all personnel (and not just the security manager) being informed about confidential errors occurring within the security infrastructure, meaning some employees might disclose sensitive information to market competitors [6].

In summary, as long as existing energy security standards do not attend to cyber security needs in areas such as home-to-grid or industrial-to-grid, it is important that smart grid technologies are made readily available by examining both common and domain-specific threats [1].

2.4. Energy Efficiency

The primary objective of any power network within the Electricity Sector is to produce reliable and stable energy to all paying customers without obstruction. Achieving such goals will require considerable amount of attentiveness, engineering effort and constant monitoring of fluctuating energy parameters. Even so, accomplishing 100% energy efficiency is still seen as close to impossible. This comes as the current centralized power systems architecture contains many technical, operational and economical limitations within its infrastructure.

On account of these limitations, the general public especially during summer periods witnesses numerous blackouts within their municipality. These blackouts are the most serious type of power outage and are triggered by numerous factors such as strong winds, flooding, fire, individuals leaving there air conditioners running for immense periods of time and even constant viewing of particular television events that occur randomly.

Because of these causes, engineers are required to anticipate the occurrence of such events and ensure that the base load is adjusted to cope with the peak in demand. This implies that a utility company must provide the least amount of electricity to its customers [2]. However, if the base load is not contemplated then a major blackout similar to that seen on August 14, 2003 which isolated the entire East Coast of the United States and Canada can be expected in the foreseeable future [2]. This disaster alone resulted in 50,000,000 customers losing power and leading to massive financial bearings [2].

Nevertheless, to tackle the issue of energy efficiency in the years to come it is fundamental that the current centralized power systems architecture be replaced by decentralized power systems architecture. This decentralized power systems architecture will have many advantages, one being that electricity can be accumulated via several storage devices located near the utility system or at an isolated site disconnected from the power grid [1].

This accumulated electricity will take immediate effect in multiple directions once a fault occurs, thus ensuring the event of a blackout will be unnoticed. Correspondingly, these storage devices will behave like a back-up generator where power is provided to the customer via renewable energy technologies and smart grid concepts. This benefit will allow engineers to repair the actual fault site without bringing about customer dissatisfaction.

3. DISTRIBUTED GENERATION SYSTEMS

Distributed Generation is also referred to as “Decentralized Generation” in parts of Asia and Europe, “Embedded Generation” across the United States and “Dispersed Generation” in North America. The technology incorporates wind turbines, micro turbines, photovoltaic systems, fuel cells, energy storage and synchronous generator applications to supply active power to distributed systems connected close to the consumers load. This section of the paper provides a brief outline of general types of distributed generation technologies and key issues facing integration of the grid.

3.1. Wind Turbines

Wind power generation harnesses the energy of wind to drive electric power generators. This is achieved using some form of wind turbine which is operated either at variable or constant speed [7]. The advantage of using wind turbines compared to many other devices comes as there is no fuel charge, no pollution, potentially a 24 hour source of energy and units are modular with fairly linear power vs. cost relationship for large-scale installations. Disadvantages include high initial cost, unpredictability of energy production and greater environmental impacts compared to solar.

Horizontal-axis wind turbines have a “yaw control” which is a mechanism to keep them pointed into the wind. This is necessary in order to maintain power production and avoid operating problems such as resonance and vibration. If the wind turbine is designed with the rotor located behind the tower, yaw adjustment will occur automatically as the wind direction shifts. As for rotors located behind the tower they create difficulties of their own including vibration, uneven speed and loss of potential given that blades pass through the wind shadow.

On the other hand, vertical-axis wind turbines have several advantages over horizontal axis wind turbines. The majority of their mechanical and electrical machinery is located on the ground rather than 40 - 90 m in the air. This makes maintenance and operation simpler and reduces structural needs. Furthermore, vertical-axis wind turbines do not need a yaw control. **Figure 3**, illustrates a variable speed induction generator connected to a rectifier and inverter.

An additional method used to operate wind turbines is by connecting the stator directly to the AC grid and the rotor through power electronic devices. **Figure 4**, illustrates this scheme cited as a doubly feed variable speed induction generator (DFIG). The doubly feed induction generator is used considerably in new wind farm installations as it offers support in power system stability and reliability during peak load conditions.

3.2. Micro-Gas Turbines

Micro-gas turbines are simple, compact and robust devices used to turn electric generators in distributed generation systems. Most micro-gas turbines take their intake air through a recuperator which is a device that manages heat from the turbines exhaust to pre-heat the intake air, raising the turbines internal temperature [3]. The recuperator is a type of radiator or heat exchanger that transfers heat from the exhaust to the incoming air. Micro-gas turbine systems are equipped with air bearings to run at speeds between 50,000 - 90,000 rpm. These systems can be mass-produced at low cost in the range of 25 - 100 kW [3]. **Figure 5**, illustrates a micro-gas turbine system before injecting a voltage into the AC grid. The generated voltage must be rectified using a diode rectifier linked to an DC/AC inverter.

3.3. Photovoltaic

Photovoltaic systems convert solar energy into electric power using an array of solar cells. Solar cells are semiconductor devices that generate DC electric power at low voltages, typically less than 0.5 Volts. A single cell may be less than a square centimeter in size and produce only a small amount of power. For this reason, many cells are connected in series to produce higher voltages and in parallel to produce higher currents. To achieve Maximum Power Point Tracking (MPPT) using photovoltaic cells a DC/DC converter must be used at the output end of **Figure 6**. This will extract the maximum available power through a given insulation level, meaning the voltage level will be maintained as close as possible to the maximum power point [8]. PV systems do not have any moving parts therefore less maintenance is required.

3.4. Fuels Cells

Fuel cells convert chemical energy directly into electrical energy. All viable fuel cell technologies consume hydrogen and oxygen. The hydrogen is obtained from a fossil fuel and the oxygen from air. Inside the hydrogen the fuel cell is oxidized under monitored and controlled conditions, in the presence of a catalyst that combines the hydrogen and oxygen to produce water. Fuel cells offer a number of advantages over rotating fossil fuel generations in distributed generations including high efficiency, low pollution, very low noise, easily re-usable heat output and quick installation [3]. Despite these tremendous advantages fuel cells provide many drawbacks including high initial costs, maintenance skills required, fuel sensitivity and unproven track record. Five types of fuel cell types include proton exchange membrane fuel cells (PEMFC), alkaline fuel cells (AFC), phosphoric acid fuel cells (PAFC), molten carbonate fuel cells (MCFC) and solid oxide fuel cells (SOFC) [3]. **Figure 7**, illustrates the basic structure of a fuel cell.

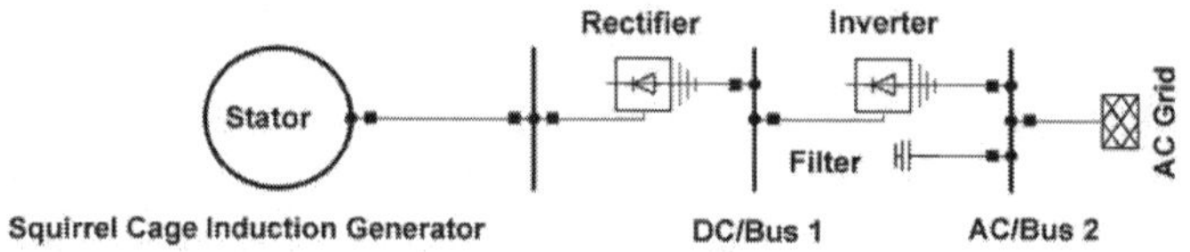

Figure 3. Variable speed induction generator.

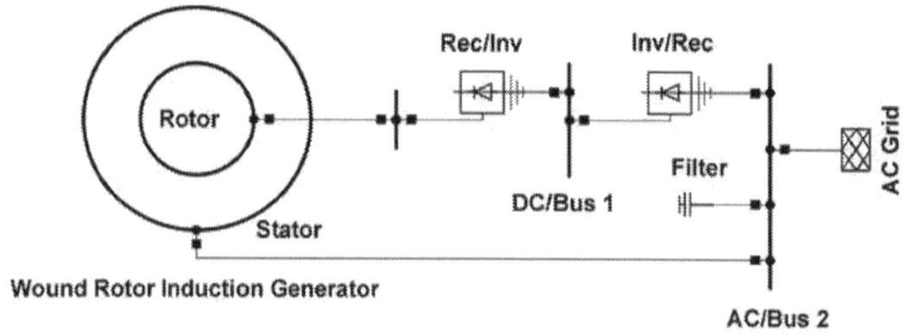

Figure 4. Double feed variable speed induction motor.

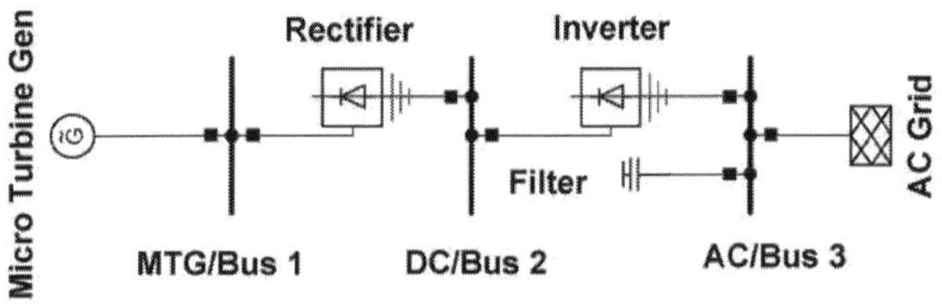

Figure 5. Micro turbine electrical system.

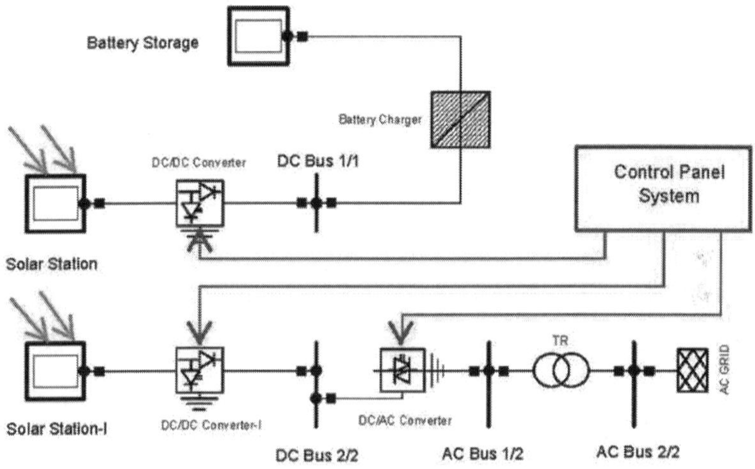

Figure 6. Photovoltaic system connected to a grid.

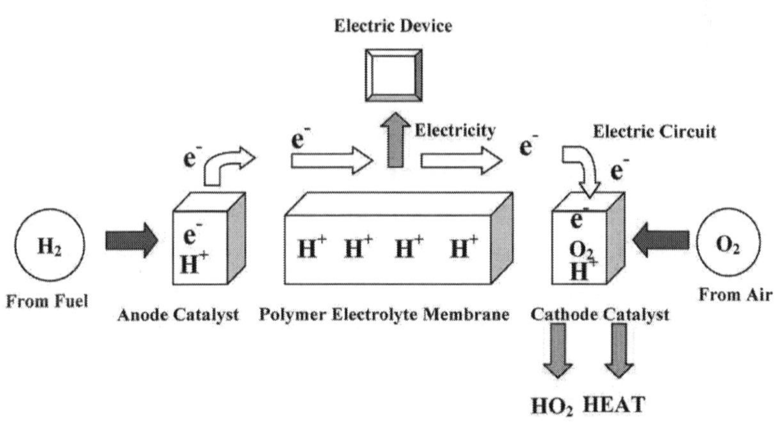

Figure 7. Fuel cell electrolyte membrane [9].

4. DISTRIBUTED GENERATION LIMITS AND IMPACTS

From a technical point of view, distributed generation systems have four limitations that must be acquired including the voltage angle (δ), voltage magnitude (V), real power (P) and reactive power (Q). These parameters can help determine the characteristics of both the PQ and PV bus. For example, in the PQ bus the voltage angle and voltage magnitude is unknown given that there is no power generation source. In the PV bus the reactive power and voltage angle are unknown, whereas the active power output and voltage magnitude are known. These buses are normally controlled by Auto Voltage Regulators (AVRs) which keep the voltage magnitude at a constant level by adjusting the field current and reactive power output. In addition, power quality also acts as a boundary given that it assists in the central generation process when overload transpires.

Despite the fact, distributed generation systems can have both positive and negative impacts on power system networks. At operational levels, impact evaluations can be analyzed through steady state analysis, dynamic analysis and protection practices [10]. Coupled distributed networks can improve system reliability, prevent under voltage drops, increase transient stability and maintain voltage levels during peak loads [11]. This occurs only if voltage levels are kept less than ±5% [12]. Conversely, negative impacts of distributed generation consist of short-circuiting behavior of components due to improper protection, anti-islanding and high fault currents [13].

5. SMART GRID TECHNOLOGIES

Smart Grid technologies are advanced electrical networks that support new generation interactive energy and communication services for the final customer. The electrical networks must be live, available, interconnected and coupled with real-time communications [14]. Smart grid technologies are the vision of the future delivering various benefits to society in order to accomplish sustainable energy development. The main advantage of smart grid implementation in terms of utility benefits include reduced perturbations and outages; minimal power losses and blackout prospects; lower maintenance and operational cost; lower greenhouse gas emissions; increased energy efficiency; increased large scale renewable energy and distributed generation integration; enabled micro-grid applications and Energy Management Systems (EMS); environmental benefits and economic growth through clean power markets. Similarly, the advantages of smart grid implementation in terms of customer benefits include enabled electricity consumption, enabled consumer services through distributed generation and cost savings through EMS. These benefits have a positive impact on society stimulating improvement in new power system networks as well as encouraging multiple engineers from pursuing research developments in smart grid technologies.

However, to achieve these benefits, the next generation power utility must co-exist with advanced grid devices and power electronic sensors. In the United State, more than half of the electricity produced is lost as a result of inefficiencies during power generation, transmission and distribution [15]. Nevertheless, smart grid technologies offer numerous characteristics to improve this state such as power quality enhancement, self-healing, generation sources, optimization of asset utilization and empowerment of users [14]. **Figure 8**, illustrates the architecture of an integrated smart grid communications platform.

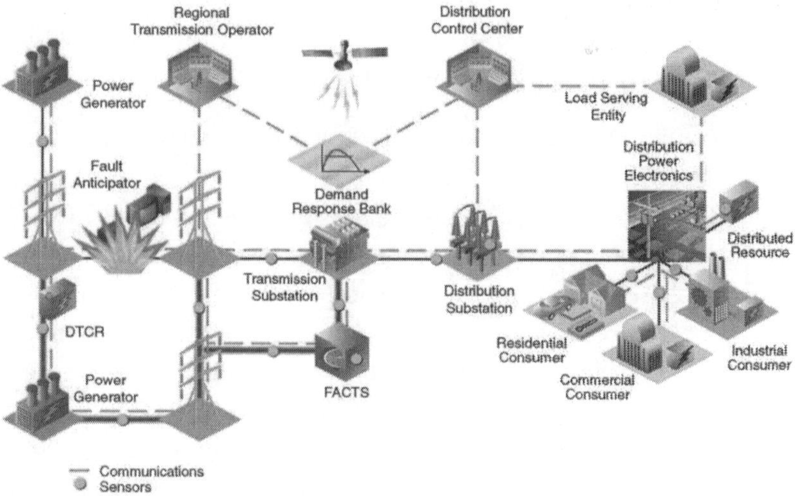

Figure 8. Architecture of integrated smart grid communications platform [16].

5.1. Continuing Research in the Field of Smart Grid Development

In the past decade, Smart Grid research has been driving developed countries to conduct more research, especially across the US and Europe. The US initially began this impetus through the Bush administration by investing resources and time to improve grid reliability, security and economics. During this period, the Department of Energy (DEO) acknowledged that the US economy lost several billions of dollars through power outages. For this reason, a policy for modernizing the national electricity was established in 2003 [17]. To achieve these goals, the American Recovery and Reinvestment Act (ARRA) introduced a $10 billion dollar investment in 2009 [18]. Several collaborative research programs were established on the basis of this incentive including the Grid Wise Alliance organization. The alliance works in coordination with the US Department of Energy and has programs on Smart Grid communication, architectures, standards, simulations, new regulations and market frameworks.

Other efforts in the US are undertaken by the Electric Power Research Institute (EPRI) through intelligent grid initiative projects [19]. EPRI deals with the technical frameworks and architectures of Smart Grid technologies by providing methods, tools and recommendations for standards to facilitate in the planning, specification and procuring of IT based systems such as advanced metering infrastructures, distributed automation, demand response management and wide-area measurement.

In contrast, the European platform on Smart Grid development aims to provide benefits for users, operators, and power companies across Europe. The platform began in 2005 and is expected to spend 750 billion Euros for improving power system infrastructures over the next 30 years. The European community has published a Strategic Research Agenda (SRA) which proposes a framework for future research themes and tasks [20]. The SRA aims to ensure that Europe's electricity network is developed in a way that enhances Europe's competitive position without damaging environmental objectives by delivering Catalyst Projects addressing potential barriers on technical standards. Europe intends to ensure electricity networks are [21]:

- Flexible—fulfilling customers needs whilst responding to changes and challenges ahead;
- Accessible—granting connection access to all network users, particularly for renewable power sources and high efficiency local generation with zero or low carbon emissions;
- Reliable—assuring and improving security and quality of supply, consistent with the demands of the digital age with resilience to hazards and uncertainties
- Economic—providing the best value through innovation, energy efficient management and "level playing field" competition and regulation.

5.2. Smart Grid Frameworks

The National Institute of Standards and Technology (NIST) [22], reveals that there are seven important Smart Grid areas that need to be addressed to make the grid "smarter". These include bulk generation, transmission, distribution, customers, operation, markets and service providers. These seven areas employ many diverse elements such as Smart Meters, Advanced Metering Infrastructures (AMI), Data Management Systems, Two-way Communication Networks, FACTS, SCADA, Digital Sensing and Distributed Automation Systems. These sections will outline each area in order to appreciate smart grid implementation.

5.2.1. Bulk Generation

Figure 9 depicts a bulk generation site that generates power from renewable and non-renewable energy sources in bulk capacity. Bulk generation is the most crucial process used to communicate, share and exchange information with market domains. Typical applications include:

- Control—Managing system power flow and reliability such as using phase angle regulators;

- Measure—Providing data collection of system performances through SCADA;
- Protect—Maintaining high quality response to faults that might cause power disruption;
- Asset Management—Identify equipment due dates for maintenance, operational history and device life expectancy;
- Record—Records data for evaluation and forecast purposes.

5.2.2. Transmission

Figure 10, reveals a transmission network that delivers electrical power to distribution networks through long distance high voltage power lines and substations. These applications require real time control devices to record and optimize data through local or wide-area networks [22]. A typical example of a transmission network incorporating smart grid technologies is that of a Phasor Measurement Unit (PMU). These units are installed to facilitate in Regional Transmission Operations (RTO) and take control during fault disturbances [23]. Current research reveals that PMUs have been effectively proven to be used for smart grid islanding and anti-islanding to integrate distributed devices such as photovoltaic (PV) panels [24].

Another transmission network that can be used is FACTS based technology like STATCOM and SVC. This provides several possibilities to enhance power system performance, power quality and maintain voltage stability [25]. SVCs can be connected directly to the grid for dynamic voltage control without the need of a step down transformer [26].

5.2.3. Distribution

Figure 11, depicts a distribution network that delivers power from a bulk generation to the customer. Most distributed networks are radial and unidirectional (i.e. during a fault scenario the operator manually restores power to the customer). This under distributed generation will become a thing of the past particularly since electricity demand continues to rise. New applications must be integrated with Advanced Distribution Management Systems (ADMS) and Distributed Power Flow Restoration (DPFR) to accomplish real time communications [27]. However, ADMS and DPFR require communication protocols between distributed networks, operations and market domains to exchange data for meter billing and energy trading. This means standards such as Distributed Network Protocols (DNP) will need to be replaced by protocols like IEC61850 seeing as the former does not support most smart grid applications.

5.2.4. Customer

Electricity customers typically are divided into commercial, industrial and residential areas. These areas consume electricity in different ways and essentially contribute to energy efficiency. **Figure 12** presents the core functionalities of smart customers where utilities are linked to Energy Service Infrastructures (ESI). As the primary bridge between end-users and Energy Management Systems (EMS), ESI provides a secure WEB portal for customers to monitor and control distributed generation, billing and on-line energy usage. This framework requires several major elements of EMS including [28]:

- Smart appliances and control panels;
- Smart meters;
- Advanced Metering Infrastructures (AMI);
- Meter Data Hubs;
- OLTP for power companies;
- Interface between Meter Data Hubs and OLTP;
- Interface between SCADA and OLTP;
- Interface between OLTP and AMI.

However, from a customer's point of view this real time Energy Management System enables end-users to receive precedented information about their electricity consumption and history, allowing them to make intelligent decisions on how to manage their electricity. The framework can only be implemented if a flexible, open and secure infrastructure is available. Zig-Bee caters for these requirements and supports the IEEE 802.15.4 standard [29]. **Figure 13**, demonstrates the Energy Management System.

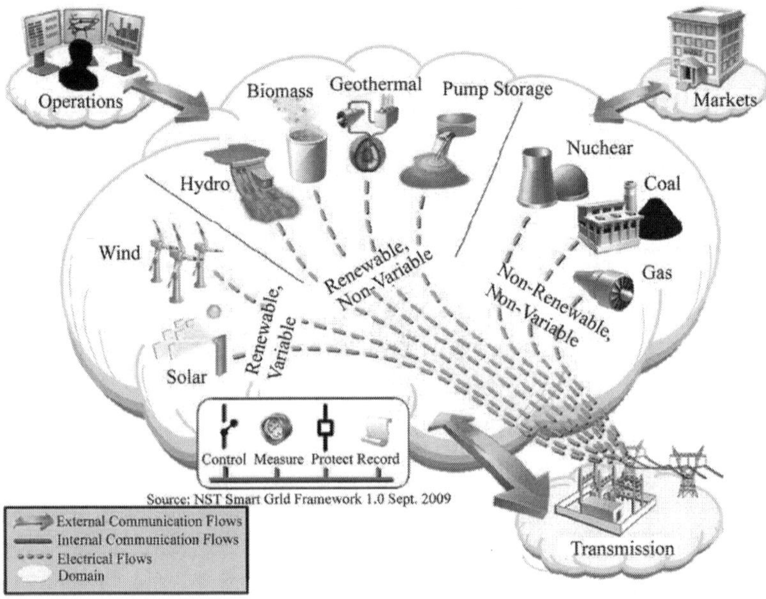

Figure 9. Bulk Generation [22].

5.2.5. Operations

Figure 14, illustrates the operations domain associated with the overall grid management. It employs the use of Energy Management Systems to ensure the network remains reliable and efficient at all times. The operations domain employs the IEC61968-1 Interface Reference Model (IRM) for applications such as [22]:

- Network Operation Monitoring;

- Local or Wide Area Control and Fault Management;
- Reporting and Statistics;
- Calculating and Analysis;
- Records and Assets Management;
- Maintenance, construction and operational planning.

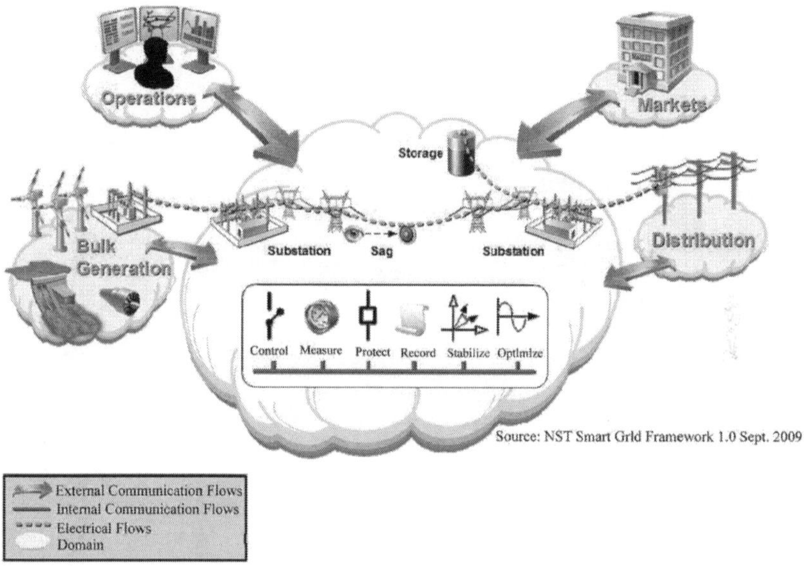

Figure 10. Transmission Networks [22].

5.2.6. Markets

Electricity cost is expected to double in the next 5-10 years due to competitive energy markets increased power demand and government regulations. However, with smart grid technologies present day costs can be minimized. **Figure 15**, illustrates the market domain responsible for managing all aspects of electricity markets such as simplification of market rules, managing the regulation and growth of trading, retailing, wholesaling and evolving electricity prices by way of consumer domains [22]. This allows for load shedding management that ultimately ensures power security of supply despite large fluctuation on the end-user.

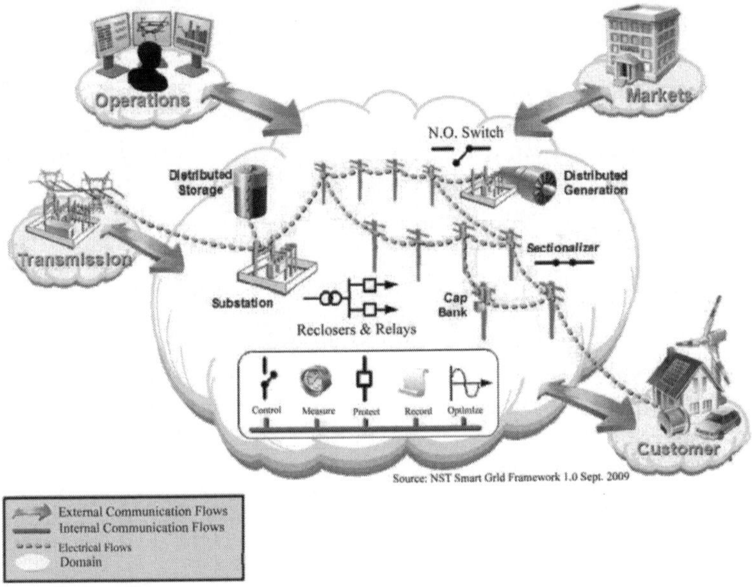

Figure 11. Distributed Networks [22].

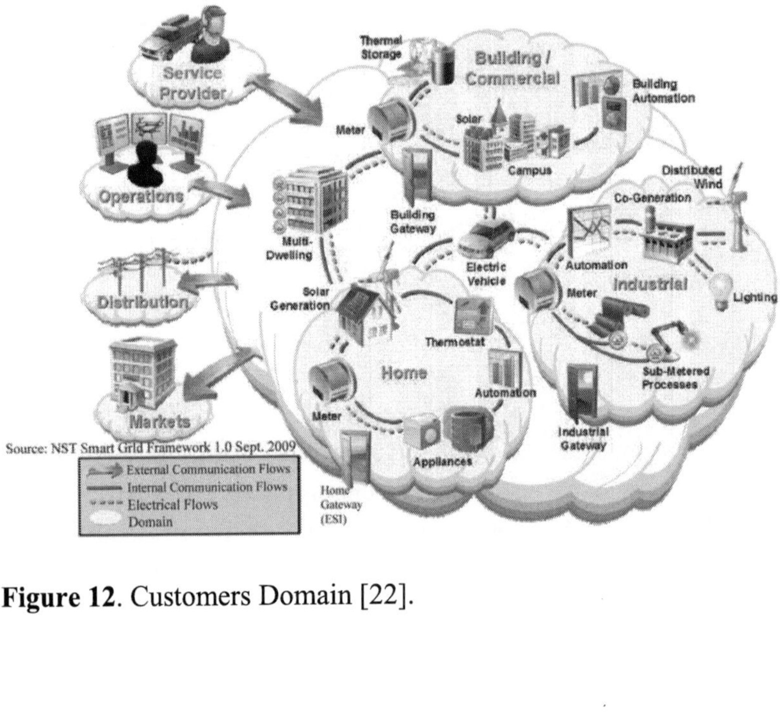

Figure 12. Customers Domain [22].

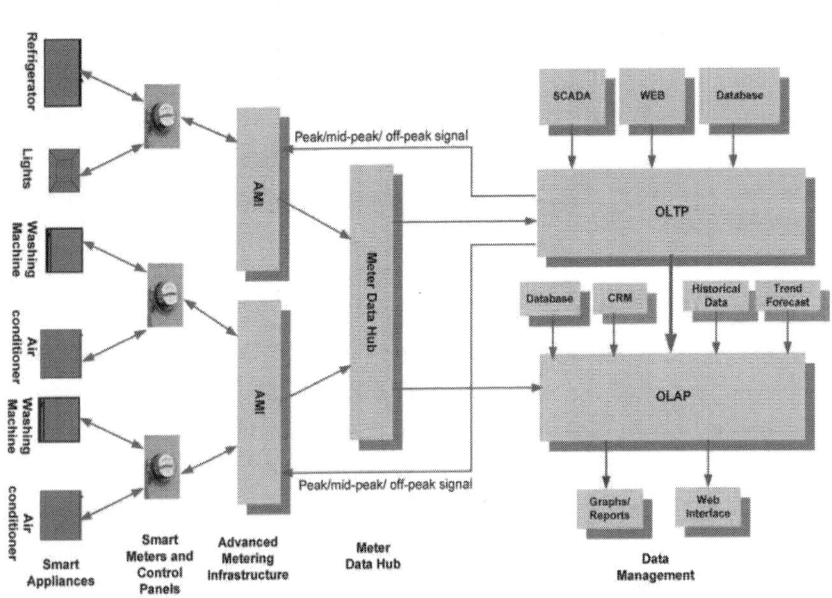

Figure 13. Real Time Energy Management System [28].

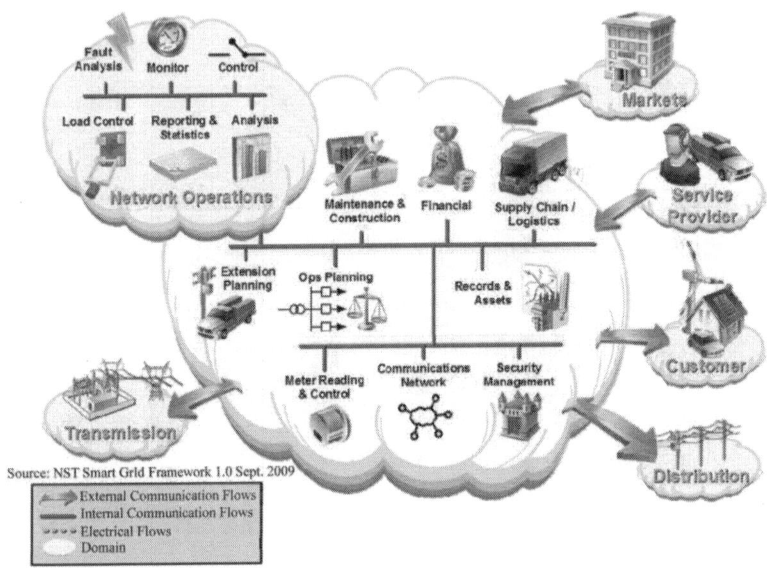

Figure 14. Operations Domain [22].

5.2.7. Service Provider

Figure 16, depicts the service provider domain comprising of three distinct domains - operations, customers and markets domains. These domains are linked to each other with extremely secure interfaces to maintain cyber security, reliability, stability and integrity when delivering services. Typical applications of service providers including managing customer installation and accounts [22]. Service providers are expected to be more innovative to end-users by offering new service offers (i.e. prepaid electricity). This model allows customer to control their energy consumption more efficient according to their pre-paid balance once inserting a digital code to their Pre-paid Advanced Metering infrastructure which converts the amount of electricity into kWh. In case of smart grid, pre-paid electricity enables the customer insight into their daily energy usage.

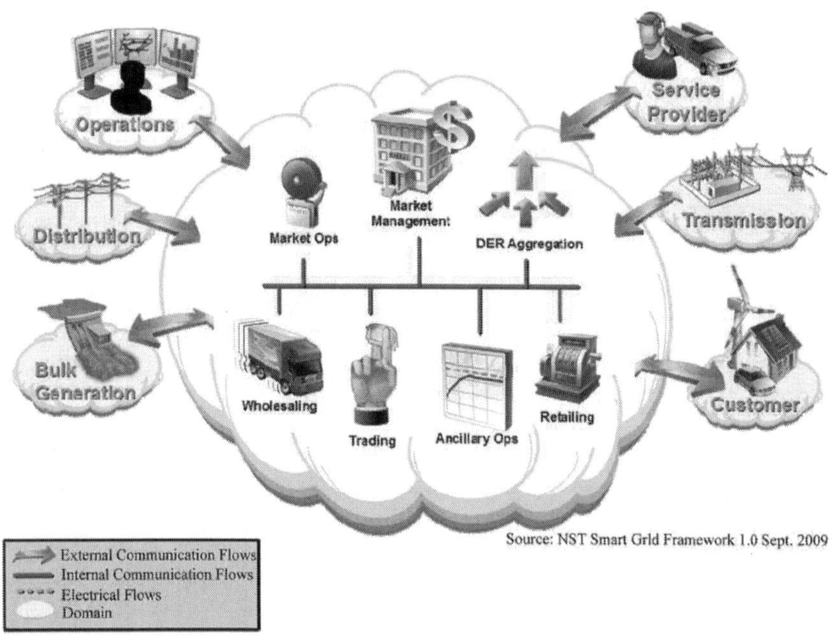

Figure 15. Markets Domain [22].

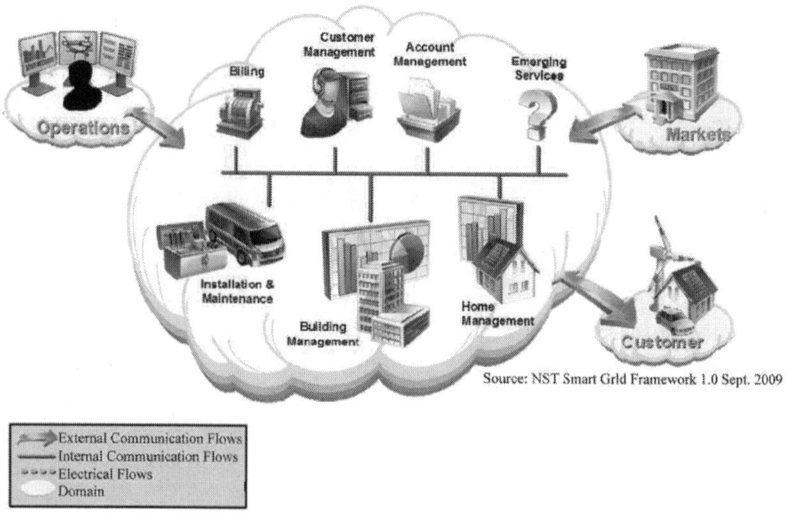

Figure 16. Service Provider Domain [22].

6. SMART GRID BARRIERS

Smart grid frameworks offer great promise for enhancing environmental compliance. However, the transition from centralized power system networks to decentralized power system networks will create major barriers when implementing smart grid technologies. Consequently, profound analysis and R&D efforts must be carried out to overcome vulnerabilities [16]. Three barriers that must be considered include technical, regulatory and economical.

6.1. Technical Barriers

Technical barriers deal with integration issues related to automation, measurement utilities, sensors, advanced power electronic devices and standard operations. Most device installations are incompatible therefore resulting in technical barriers. Interoperability is the ability to share information that has been exchanged between two or more systems [30]. These systems may consist of Power Quality Monitoring (PQM), Wide Area Measurement Systems (WAMS), SCADA and various intelligent controls.

6.2. Regulatory Barriers

On account of the lack of international standards, interoperability methods of different products are distinct in different localities. For example, many states regulate their standards, rates and policies based on local regulations. These conditions can slow down smart grid development and even increase the cost of integration. However upgrading power system networks with smart grid technologies will allow the consumer, service provider and operator to share benefits collectively. This advantage is difficult to measure from a financial point of view, guaranteeing regulators address the operational standards based on wholesale market liberalization.

6.3. Economical Barriers

Smart grid technologies offer various benefits for utilities and consumers to manage their energy consumption. Smart grids provide flexibility for new integration of distributed generation. However, the implementation of smart grid technologies requires significant investment to upgrade the existing power system infrastructure as many devices are still expensive. Furthermore, smart grid applications moreover allow selling of small amounts of energy in liberalized market.

7. CONCLUSIONS

The authors of this paper evaluated some of the key debatable issues influencing the Electricity Sector both at present and in the foreseeable future. Following such assessment research on distributed generation systems and smart grid technologies was emphasized concentrating on previous R&D efforts across US and Europe. In addition, the paper provided a comprehensive analysis into seven smart grid frameworks that need to be addressed to make the future electricity grid smarter. These include bulk generation, transmission, distribution, customer, operation, markets and service providers. Finally, a brief impression on smart grid barriers including technical, regulatory and economical was discussed. Positive and negative impacts were also mentioned including improving system reliability, preventing under voltage drop, influencing transient stability and harmonic distortion during connection or disconnection of these technologies.

Distributed Generation Systems, Smart Grid Technologies and Future Motivators Influencing Change in the Electricity Sector are currently an important issue. Therefore, since it has become a leading concern in the field of power system in particular with the development in the industrial nations worldwide, this paper will greatly valuable for further understanding in these subjects.

ACKNOWLEDGEMENTS

The first author would like to acknowledge the Australian Development Scholarship (ADS) for providing the support essential to pursue his Master by Research at Victoria University, Melbourne, Australia.

REFERENCES

1 B. Stojcevski and A. Kalam, "Fault Location in Overhead Power Lines Using the IEC61850 International Protocol," International Review on Modelling and Simulation, vol. 3, No. 5, 2010, pp. 888-899.

2 R. Garnaut, "The Garnaut Climate Change Review," Cambridge University Press, Port Melbourne, 2008.

3 H. L. Willis and W. G. Scott, "Distributed Power Generation—Planning and Evalutation," Marcel Dekker, New York, 2000.

4 R. Whitaker, "Understanding Climate Change—The Story of the Century", New Holland Publishers, Sydney, 2007.

5 A. Jolly, "Managing Climate Risk—A practical," Thorogood, London, 2008.

6 A. S. Massoud, "Electricity infrastructure security: Toward reliable, resilient and secure cyber-physical power and energy systems,"

Power and Energy Society General Meeting, 2010 IEEE, Minneapolis, 30 September 2010.

7 V. P. Mahadanaarachchi, R. Ramakuma, "Impact of distributed generation on distance protection performance—A review," Power and Energy Society General Meeting—Conversion and Delivery of Electrical Energy in the 21st Century, 2008 IEEE, Pittsburgh, 20-24 July 2008, pp.1-7.

8 N. Hatziargyriou, "Modeling New Forms of Generation and Storage," CIGRE Technical brochure, TF 38.01.10, November 2000.

9 W. El-khattam and M. M. A. Salama, "Distributed Generation Technologies, definitions and benefits," Electric Power Systems Research, Vol. 71, 2004, pp. 119-12.

10 J. A. Lopes, "Integration of Dispersed Generation on Distribution Networks-Impact Studies," IEEE, Vol. 1, 2002, pp. 323-328.

11 R. A. Prata, "Impact of Distributed Generation Connection with Distribution Grids—Two Case-Studies," IEEE Power Engineering Society General Meeting, Pittsburg, 2006, p. 8.

12 A. Mahmood, M. Aamir and M. I. Anis, "Design and Implementation of AMR Smart Grid System," IEEE Electric Power and Energy Conference, Canada, 2008, pp. 1-6.

13 C. Chai, L. Wei-Jen, P. Fuangfoo, M. Williams and J. R. Liao, "System Impact Study for the Interconnection of Wind Generation and Utility System," IEEE Transactions on Industry Applications, vol. 41, no. 1, 2005, pp. 163- 168.

14 J. R. Roncero, "Integration is Key to Smart Grid Management," Smart Grids for Distribution, 2008. IET-CIRED. CIRED Seminar, Frankfurt, 23-24 June 2008, pp. 1-4.

15 T. J. Lui, W. Stirling and H. O. Marcy, "Get Smart," Power and Energy Magazine, IEEE, vol. 8, no. 3, MayJune 2010, pp. 66-78.

16 C. W. Gellings, M. Samotyj and B. Howe, "The Future's Smart Delivery System [Electric Power Supply]," Power and Energy Magazine, IEEE, vol. 2, no. 5, SeptemberOctober 2004, pp. 40-48.

17 "Grid 2030—A National Vision for Electricity's Second 100 Years," U.S Department of Energy, Office of Electric Transmission and Distribution, July 2003 Available from HUhttp://www.oe. energy. gov/DocumentsandMedia/Electric_Vision_Document.pdf

18 National Institute of Standart and Technology, "NIST and the Smart Grid," 2010. http://www.nist.gov/smartgrid/nistandsmartgrid.cfm

19 Electric Power Research Institute, "Intelligrid," 2010. http:// intelligrid.epri.com

20 European Commission, Directorate General for Research, Directorate Energy, "Strategic Research Agenda for Europe's Electricity Networks of the Future," European Technology Platform Smart Grid, 2007.

21 European Commission, "European Smart Grid Technology Platform—Vision and Strategy for European Electricity Networks of the future," 2010. http://www.smartgrids.eu/web/node/28/default.asp

22 Office of the National Coordinator for Smart Grid Interoperability, National Institute of Standart and Technology, U.S. Department of Commerce, "NIST Framework and Roadmap for Smart Grid Interoperability Standard, Realease 1.0," NIST Special Publication 1108 on the January 2010.

23 J. de La Ree, V. Centeno, J. S. Thorp and A. G. Phadke, "Syncrhonized Phasor Measurement Applications in Power Systems," IEEE Trans. On Smart Grid. vol. 1, no. 1, June 2010, pp. 20-27.

24 Schewizer Engineering Laboratories, "Effectively Integrate More Photovoltaic (PV) Generation Sources into the Grid," 2010. https://tesla.selinc.com/emailmarketer/display.php?M=12626&C=71 14023c0fe9099bf0e911a246630d60&S=250&L=7&N=216U

25 N. A. Hidayatullah, Z. J. Paracha and A. Kalam, "Impacts of Distributed Generation on Smart Grid," International Conference of Electrical Energy and Industrial Electronic System (EEIES), Malaysia, 7-8 December 2009.

26 R. Griinbaum, M. Halonen and G. Stromberg, "SVC for 69 kV Direct Grid Connection," Transmission and Distribution Conference and Exposition, 2008. T&D. IEEE/ PES, Chicago, 21-24 April 2008, pp. 1-7.

27 J. Fan and S. Borlase, "The Evolution of Distribution," IEEE Power & Energy Magazine, vol. 7, no. 2, MarchApril 2009, pp. 63-68.

28 D. Y. R. Nagesh, J. V. V. Krishna and S. S. Tulasiram, "A Real-Time Architecture for Smart Energy Management," Innovative Smart Grid Technologies (ISGT), Gaithersburg, 19-21 January 2010, pp.1-4.

29 J.-S. Lee and Y.-C. Huang, "ITRI ZBnode: A ZigBee/IEEE 802.15.4 Platform for Wireless Sensor Networks," Systems, Man and Cybernetics, 2006. SMC '06. IEEE International Conference on, Taipei, 8-11 October 2006, pp. 1462-1467.

30 The GridWise Architecture Council, "GridWise□ Interoperability Context-Setting Framework," 2010. http://www. gridwiseac. org/ pdfs/interopframework_v1_1.pdf

CHAPTER 4

Influence of Control Modes of Grid-Connected Solar Photovoltaic Generation on Grid Power Flow

Zakaria Al-Omari

Electrical Engineering Department, Faculty of Engineering,
Al-Ahliyya Amman University, Amman, Jordan

ABSTRACT

Integration of Solar Photovoltaic (PV) generation into an existing distribution system has many impacts on the system, with the power flow being one of the major issues. This impact is not ge- neric for any network, but it may manifest itself either positively or negatively, depending on the grid configuration, interface control modes, operation mode, and load profile. Grid-connected PV systems have three control options of the local voltage controller of the interface DC-AC converter. These control modes are Power Factor control, voltage control, and Droop Voltage control. This paper aims at evaluating and comparing the impacts of those control modes on the grid power flow. A set of evaluation criteria and indices is defined and mathematically formulated. Based on the requirements of the used program (Power Factory Dig

Silent V14.1.3), a computation plan (algorithm) has been proposed. The algorithm has been applied to a typical weak network and a wide range of simulations has been carried out. Simulation results have been thoroughly dis- cussed and important findings have been concluded.

KEYWORDS

Grid-Connected PV, Control Modes, Power Flow, Power Factory Dig Silent Software

1. INTRODUCTION

For nearly a century, most countries have relied on just one model of power distribution: sending electricity over huge transmission grids from big central generating plants to customers in their homes, offices and factories [1]. A parallel trend, however, has been developing in recent years, in which significantly smaller sized generation units are being connected at the distribution level. This technology is called Distributed Generation (DG). The portfolio of DG technologies includes reciprocating engines, micro turbines, combustion gas turbines (including mini turbines), solar thermal, fuel cells, wind turbines and Solar Photovoltaics (PV). Recent studies suggest that in medium and long terms, PV generator will become commercially so attractive that large-scale implementation of this type can be seen in many parts of the world [2] [3] .

Due to the significant advantages of grid-connected PV systems, this paper focuses on this type of DG tech- nology. Grid-connected PV systems

- will reduce the power bill as it is possible to sell surplus electricity produced to the local electricity supplier, comparatively easier to install as they do not require a battery system.
- are comparatively easier to install as they do not require a battery system.

- have the advantage of effective utilization of generated power because there are no storage losses involved.
- are growing rapidly due to accelerated cost reductions and associated growth in production. As these tech- nologies mature they have the potential to provide an increasing share of electricity demand.

However, integration of Solar Photovoltaic (PV) generation into an existing distribution system has many im- pacts on the system, with the power flow being one of the major issues. This impact is not generic for any net- work, but it may manifest itself either positively or negatively, depending on the grid configuration, interface control modes, operation mode, and load profile [4] . Grid-connected PV systems have three control options of the local voltage controller of the interface DC-AC converter. These control modes are Power Factor (PF) con- trol, Voltage (V) control, and Droop Voltage (DV) control.

The paper quantifies the influence of the three control options on the bus voltages and power losses of the host distribution network. For achieving this objective the power flow approach has been used, which is consi- dered as the optimal tool for analyzing the steady state operation of the network.

2. POWER FLOW IN ACTIVE DISTRIBUTION NETWORKS

2.1. Overview

Modern distribution systems were designed to accept bulk power at the bulk supply transformers and to distri- bute it to customers. Thus the flow of both real power (P) and reactive power (Q) was always from the higher to the lower voltage levels. However, with PV generators, the distribution network is no longer a passive circuit supplying loads but an active system with power flows and voltages determined by the generation as well as the loads. The voltage source convertor of the PV generator will allow export of real and reactive power but may introduce harmonic currents. Thus the power flows through the circuits may be in either direction depend- ing on the relative magnitudes of the real and reactive network loads compared to the generator outputs and any losses in the network. The change in real and reactive power flows caused by PV generators has important tech- nical and economic implications for the power system. However, additional operational costs can be avoided through optimal correction in electrical systems [5] .

The power flow approach in distribution networks hosting PV generators differs from the conventional one mainly in two issues; the selection process of the slack bus and the modelling of the generation buses. In a transmission network, a bus with a large generating unit is usually chosen as the slack bus. In a distribution network, the bus where the distribution system under investigation is connected to a higher voltage level is usually chosen as the slack bus. The complex voltage is specified at slack bus; the voltage angle is usually set to zero [6] .

2.2. Operation Modes of a PV Generator

The PV generator bus is modelled through one of three models depending on the control mode of the converter. Three control modes are considered [7] [8] :

- Power Factor (PF) Control

The PF control of the converter is enabled, i.e. the converter operates at constant power factor. The used bus model is PQ type, i.e. the injected P (P_g) and injected Q (Q_g) are specified.

$$\frac{P_g}{\sqrt{P_g^2 + Q_g^2}} = \cos\theta \qquad (1)$$

where $\cos\theta$ is the set PF.

This operation mode has two constraints: the rated MVA of the converter may not be violated, and the PF should be maintained constant.

- Voltage Control

The voltage control of the converter is enabled; that means, the converter operates at variable power factor operation. The used bus model is PV type, i.e. the injected P and the voltage magnitude are specified. This operation mode allows the generator to support the voltage profile in the network through injecting reactive power. Voltage control can be done locally, i.e. the reactive power output of the generator is controlled to achieve the specified local voltage at its terminal. The active power output is constant for the dispatch.

- Droop Voltage Control

For this control mode the voltage is controlled according to a specified droop (e.g. droop = 2%) and set vol- tage. The droop is defined as:

$$\text{droop}\% = \frac{\Delta V / V_n}{\Delta Q / Q_n} \times 100 \tag{2}$$

The generation bus model for this mode is DV type, which is new for power flow calculations (i.e. not in- cluded in the conventional power flow). The generator can be set to control the local voltage at its terminal to a specified set point. With droop control the set point is not reached in any case because the set point is moved (by DV droop) as more reactive power is needed to reach the original voltage set point of the PV generator. The ad- vantage of the droop control is that more than one machine at one bus could control the voltage as well as the participation of the single machine could be configured with the setting of the droop value. When set to voltage control, a droop value can be entered. The voltage at the local bus is then controlled according to the following equations [7] :

$$V = V_{\text{setpoint}} - dV_{\text{drop}} \tag{3a}$$

$$dV_{drop} = \frac{Q - Q_{setoint}}{Q_{drop}} \tag{3b}$$

$$Q_{droop} = \frac{S_{norm} \times 100}{droop} \tag{3c}$$

where

V is the actual voltage value at the terminal busbar.

$V_{setpoint}$ is the specified voltage set point of the PV generator.

Q is the actual reactive power output of the PV generator.

$Q_{setpoint}$ is the specified dispatch reactive power of the PV generator.

S_{norm} is the nominal apparent power.

droop is the droop value specified in percentage.

3. EVALUATION CRITERIA (INDICES)

For achieving a systematic methodology to quantify the required impact, a set of evaluation criteria and indices are defined and mathematically formulated.

3.1. Penetration Index (Penetration Level)

The penetration level of distributed generation (DG) on the distribution network is measured against total load demand or the percentage of DG power referred to the rated power of the network [8] -[11] . It is defined as the ratio of the PV capacity to the peak load of the hosting distribution network

$$\mathbf{PI\%\ \ or\ \ PL\%} = \frac{\mathbf{P_{PV}}}{\mathbf{P_{Load}}} \times 100 \tag{4}$$

3.2. Dispersion Index (Dispersion Level) [8]

It is defined as the ratio of the number of PV buses to the number of load buses:

$$\mathbf{DI\ \ or\ \ DL\%} = \frac{\mathbf{\#PV - Buses}}{\mathbf{\#Load - Buses}} \times 100 \tag{5}$$

This index points out to the geographic dispersion of PV generators and not its capacity.

3.3. Voltage Rise Index [6]

Voltage rise index (VRI%) for a bus is the ratio of the difference between bus voltage with PV generation and bus voltage without PV generation, to the bus voltage without PV generation:

$$\mathbf{VRI\ \%} = \frac{V_{\mathbf{+PV}} - V_{\mathbf{-PV}}}{V_{\mathbf{-PV}}} \times 100 \tag{6}$$

The German VDEW code allows VRI% = 2% as a maximum. This figure is suitable for strong grids. But for weak grids, this index is useful only as an indicator for the voltage support degree by PV. Therefore I suggest an alternative index $VRI_2\%$, which indicates the voltage rise relative to the nominal voltage:

$$\mathbf{VRI_2\%} = \frac{V_{+PV} - V_n}{V_n} \times 100 \tag{7}$$

3.4. Total Loss Index [6]

It is defined as the ratio of the power loss (P or Q) in the active host grid to the power loss in the passive host grid:

$$\mathbf{TLI\ \%(P\ or\ Q)} = \frac{\mathbf{P\ or\ Q\ Loss_{+PV}}}{\mathbf{P\ or\ Q\ Loss_{-PV}}} \times 100 \tag{8}$$

This index shows the reduction or increase of total losses caused by PV inclusion relative to the original losses, but it gives no information about the loss rate relative to the input power (P_i or Q_i). Therefore further power loss indices are introduced, referred to as ΔPI% and ΔQI%, and defined as:

$$\mathbf{\Delta PI\ \% = P\ Loss\ \%} = \frac{\mathbf{P\ Loss}}{P_i} \times 100 \tag{9a}$$

$$\mathbf{\Delta QI\ \% = Q\ Loss\ \%} = \frac{\mathbf{Q\ Loss}}{Q_i} \times 100 \tag{9b}$$

4. TEST NETWORK

Several simulations have been carried out using a typical distribution grid with the voltage level 20/0.4 kV. The test network is shown in Figure 1 with several PV units, which can be connected or disconnected to the grid through corresponding circuit breakers. The load profile has been determined so that the passive grid (base case) represents a weak network, Viz. a network with a bad voltage profile. The test network consists of 12 buses, 9 lines, 9 distributed line loads and 4 concentrated loads at the critical buses 2, 8, 9, and 12.

5. THE ALGORITHM

Based on the mathematical models and the requirements of the used program Dig Silent Power Factory (Dig Si- lent-PF) V14.1.3, which is a computer aided engineering tool for the analysis of electrical power systems, an algorithm has been developed. The algorithm is designed so that it can be applied to any active distribution grid with multiple voltage levels and different PV configurations. This algorithm will be adopted to investigate the test network. The developed approach is shown as a flow chart in Figure 2.

6. COMPUTER SIMULATIONS

6.1. Based Case

Initially a power flow was carried out for the distribution network without PVs (Base Case). The results show that the network has two problems: Significant total P and Q losses, whose ratios are 6.45% and 8.3% re- spectively, and under voltage of the half of buses (Figure 3).

To evaluate the impact of PV on power flow of base case, several scenarios was simulated. These scenarios include connecting PV units to the grid with different capacities (i.e. different PL%) and locations (i.e. different DI%).

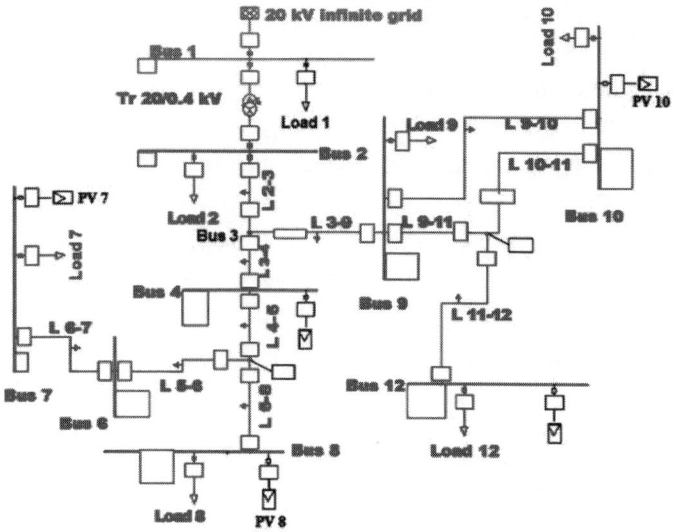

Figure 1. The test network.

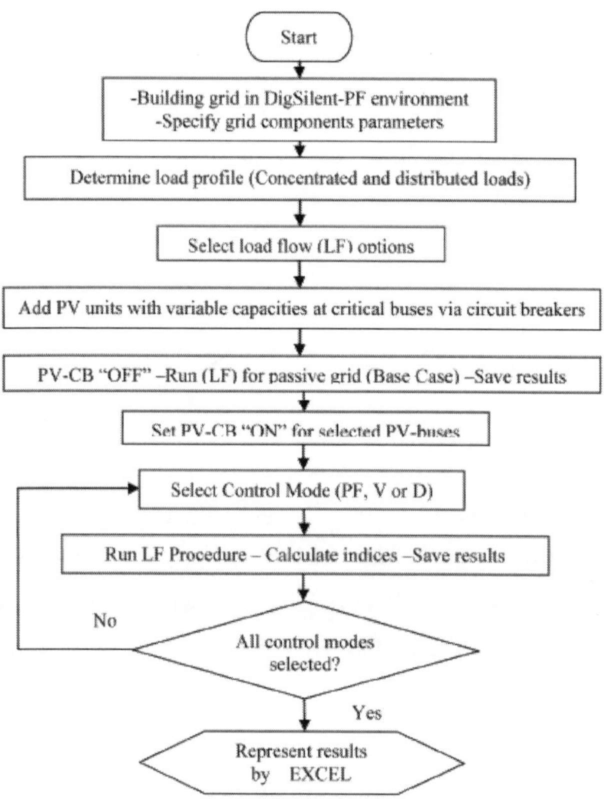

Figure 2. Algorithm flowchart.

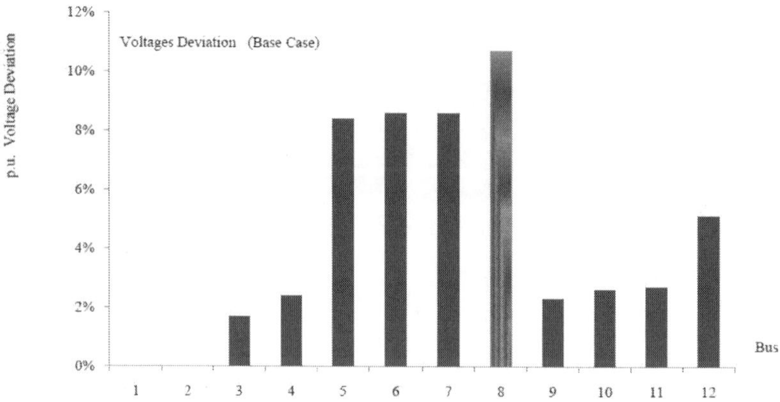

Figure 3. Bus voltage of the network for base case.

6.2. Connecting a 200 KVA-PV at Bus 8

The connection of a single PV generator at bus 8 with the three control modes has, in general, improved the network performance, but to an extent that is depending on the control mode. The major impacts are:

- The voltage profile at all buses has been improved and became acceptable (Figure 4).

- The power flows are reversed on several lines.

- Figure 5 shows voltage rise index (VRI%) of network buses for the three control modes. It is clear that the PF control gives the highest VRI with an average value of 10.2% for the buses 5, 6, and 7, which are adjacent to the PV connection bus 8. The Droop voltage control gives the next best average value of VRI (8.2%), while the voltage control gives the lowest VRI (7.7%).

- The P losses index $\Delta PI\,\%$ has been reduced to 3.23% for the PF control and the droop voltage control and unchanged for V control, while Q losses index $\Delta QI\,\%$ increased for V control to 10.5% while remained unchanged as for the base case for DV and PF control modes (Figure 6).

6.3. Connecting 300 kVA PV at Bus 8

A PV generator of a much more capacity (300 kVA) is connected to the same bus 8. The simulation results show the following impacts:

Ø The impact on bus voltages are shown in Figure 7. The voltage profile has been well improved. Again, the PF control gives the highest voltages. But due to the higher penetration level of this scenario, the voltages of at least 3 buses, namely the PV connection bus 8 (1.1 pu), and the close adjacent buses 5, 6 and 7, violate the upper limit (1.05 pu). Even, the voltage of bus 8 violates the 1.05 limit for Droop control. Therefore, it is recommended to determine the maximum PV capacity allowed to be added to the bus. In this case study, the maximum allowed PV capacity was found to be 230 kVA.

Ø Concerning the P and Q losses indices ($\Delta PI\,\%$ and $\Delta QI\,\%$), adding the 300 kVA PV causes the follow- ing impacts relative to the base case of the network (Figure 8):

- For PF control, $\Delta QI\,\%$ is unchanged, while $\Delta PI\,\%$ increases by about 30% relative to base case.

- For both V and DV control, $\Delta QI\,\%$ becomes about 3 times and 2 times respectively relative to that of base case, while $\Delta PI\,\%$ becomes 5 times and 3 times respectively.

These negative impacts on voltages and power losses occurred mainly because of the high penetration level (about 100%); so it is recommended to avoid that as it is explained in the next section.

6.4. Impact of Penetration Level on Power Flow

With the interconnection location being fixed (i.e. DI% is fixed), the impacts of PV with PF control mode on power flow with different capacities are investigated. Figure 9 shows the impact of different PL on the power loss in the host grid. From the trend lines plotted on the basis of simulation results for PF control mode, it is ob- vious that for low PV penetration levels P and Q losses decrease reaching a minimum for almost PI = 30% and 70% respectively, while losses increase for higher penetration levels and particularly P losses reach val- ues that are much higher than losses of base case from PI of 55% onwards.

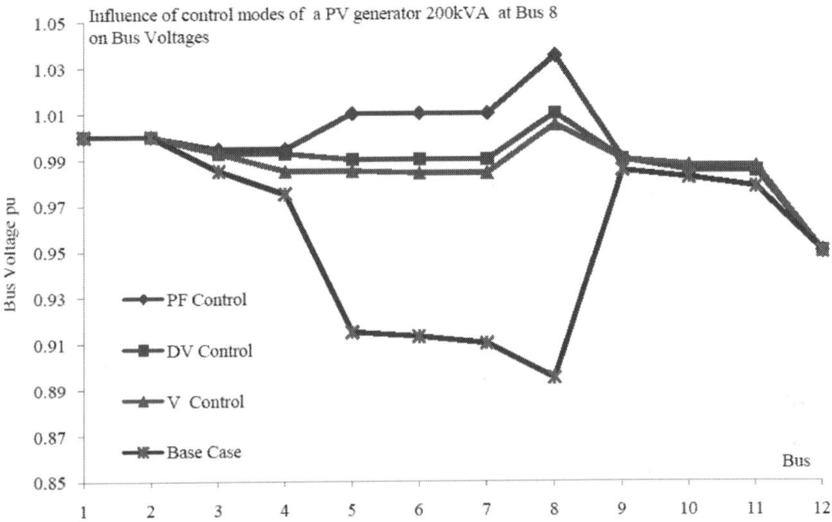

Figure 4. Bus voltage for control modes of a 200 kVA PV generation at bus 8.

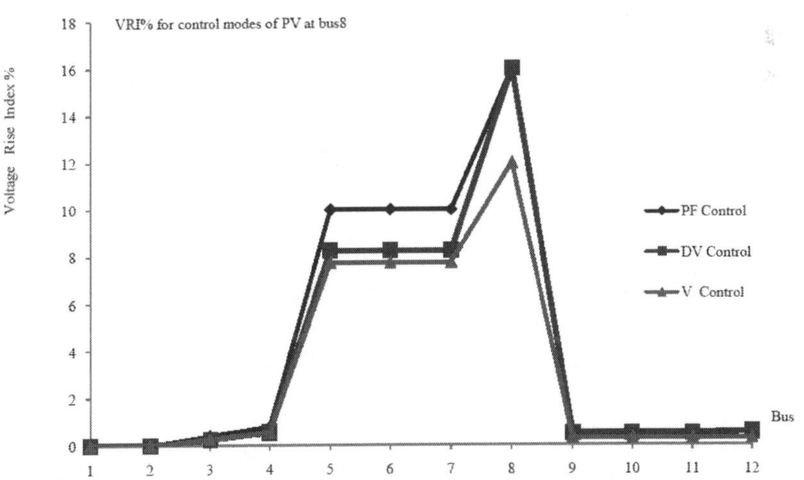

Figure 5. Voltage rise index of buses caused by 200 kVA PV at bus 8.

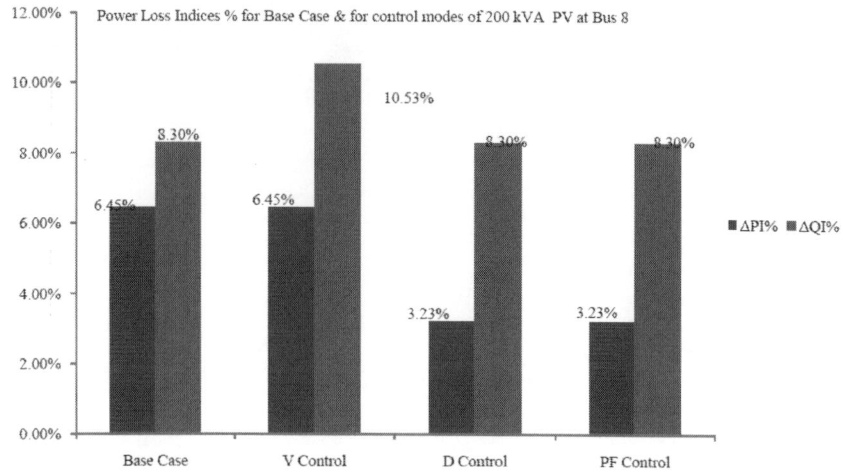

Figure 6. Power loss indices (ΔPI% and ΔQI%) for 200 kVA PV at bus 8.

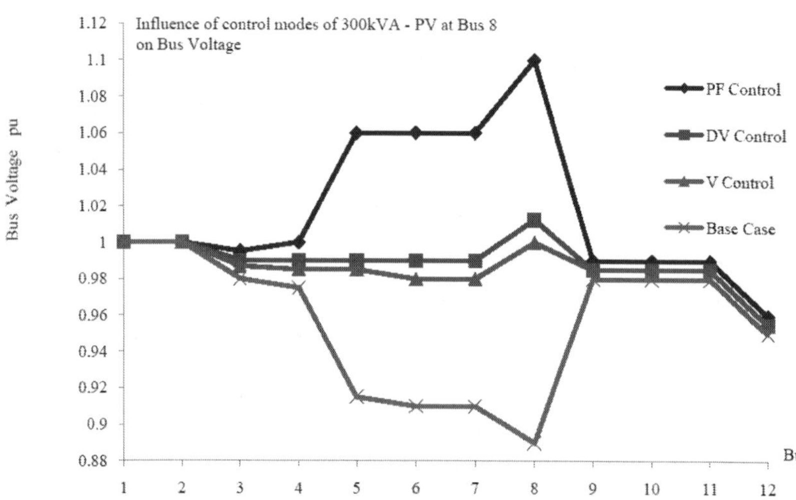

Figure 7. Bus voltages for 3 control modes of 300 kVA PV at bus 8.

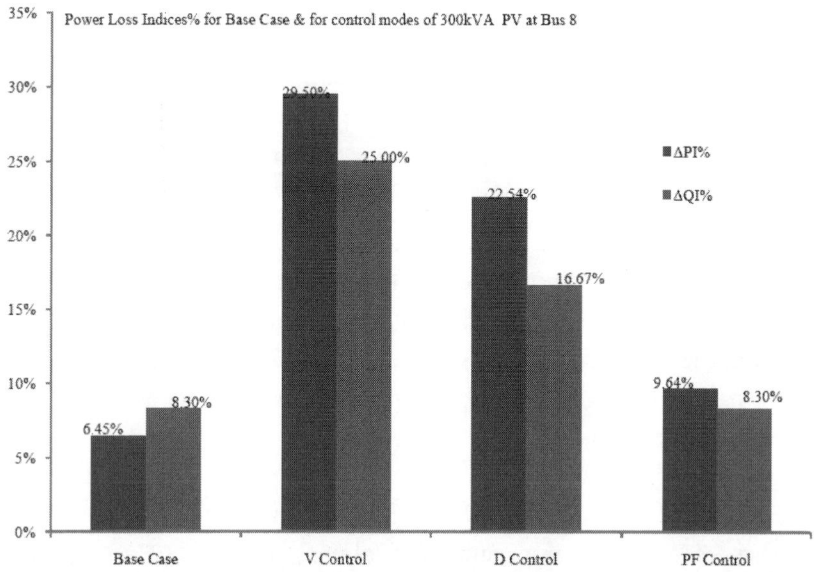

Figure 8. Power loss indices for 3 control modes of 300 kVA PV at bus 8.

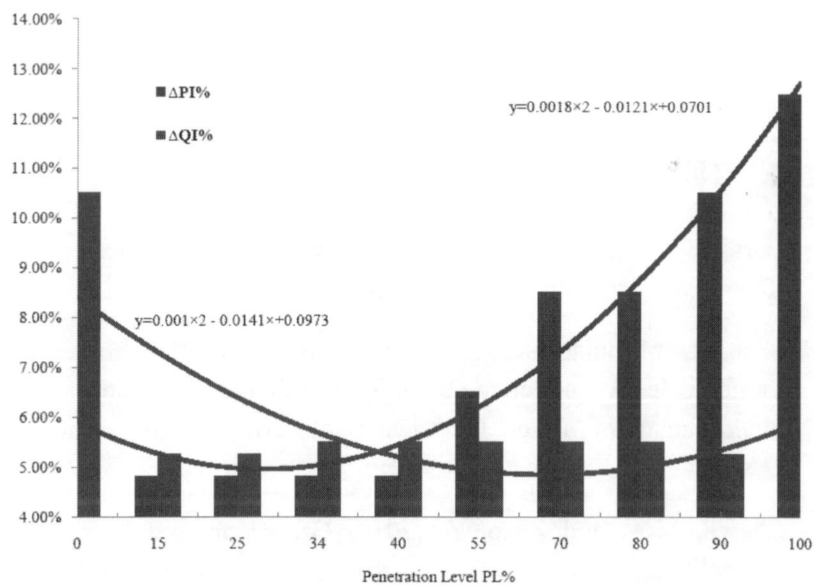

Figure 9. Power loss indices as a function of penetration level for PF control.

7. CONCLUSIONS

The simulation results show that the connection of a single PV generator at a bus with the three control modes has, in general, improved the network performance, but to an extent that is depending on the control mode and penetration level (PL%). The power factor (PF) control gives the highest voltage rise index (VRI) for buses which are adjacent to the PV connection bus. The Droop voltage control gives the next best average value of VRI, while the voltage control gives the lowest VRI. With a higher penetration level (about 60% - 70% PL), the PF control causes that the voltages of the PV connection bus and its adjacent buses violate the upper voltage limit.

For PF control mode with low PV penetration levels, P and Q losses decrease reaching a minimum for almost PL of 30% and 70% respectively, while losses increase for higher penetration levels, and particularly P losses reach values that are much higher than losses of base case from PL of 55% onwards.

It is recommended to determine the maximum PV capacity allowed to be added to the desired bus for the concerned network, in order to avoid the negative impacts on the network performance.

REFERENCES

1 Gardiner, B. (2013) Bypassing the Power Grid Published. Special Report October 8.http://www.nytimes.com/

2 Shah, R., Mithulananathan, N. and Bansal, R. (2012) Influence of Large-Scale PV on Voltage Stability of Sub-Trans- mission System. International Journal on Electrical Engineering and Informatics, 4, 148-161.

3 Eltawil, M.A. and Zhao, Z. (2010) Grid-Connected Photovoltaic Power Systems: Technical and Potential Problems. Review of Renewable and Sustainable Energy, 14, 112-129.

4 Thong, V.V., Vermeyen, P., Driesen, J. and Belmans, R. (2004) Grid Connection Aspects of Distributed Generation. Pro- ceedings of UIE International Conference, Durban, 18-23 January 2004, 14.

5 Al-Omari, Z. and Abdallah, J. (2008) Modeling Additional Operational Costs Incurred Due to Absent of the Optimal Correction in Electrical Systems. Journal of Applied Sciences, 8, 4422-2227.

6 Hamzeh, A. Integration of Distribution Generation into Electrical Distribution Generation, © 2010-2011. Damascus University Publications, Damascus.

7 DIgSILENT GmbH, Germany (2011) Technical Documentation for Power Factory®Simulation Tool. Copyright 2011. http:// www. digsilent.de/

8 González-Longatt, F.M. (2007) Impact of Distributed Generation over Power Losses on Distribution System. Pro- ceedings of 9th International Conference Electrical Power Quality and Utilization, Loughborough University, UK, 2007, 9-11.

9 Balamurugana, K., Srinivasana, D. and Reindlb, T. (2012) Impact of Distributed Generation on Power Distribution Sys- tems. Energy Procedia, 25, 93-100.

10 Canova, A., Giaccone, L., Spertino, F. and Tartaglia, M. (2009) Electrical Impact of Photovoltaic Plant in Distributed Net- work. IEEE Transactions on Industry Applications, 45, 341-347.

11 Sheikhi, A., Maani, A., Safe, F. and Ranjbar, A.M. (2013) Distributed Generation Penetration Impact on Distribution Networks Loss. Proceeding of International Conference on Renewable Energies and Power Quality (ICREPQ'13), Bilbao, 20-22 March 2013.

CHAPTER 5

Impact of Distributed Generation on Smart Grid Transient Stability

Nur Asyik Hidayatullah, Zahir J. Paracha, Akhtar Kalam

School of Engineering and Science, Victoria University, Melbourne, Australia.

ABSTRACT

In the 21st century Smart Grid and Renewable Energy technologies are an important issue with regards to global climate change problem and energy security. The evolution of current conventional or centralized generation in form of distributed generation and Smart Power Grid (SPG) has great opportunity and potentially can eradicate several issues associated with energy efficiency, energy security and the drawback of aging power system infrastructures. In order to meet the rising electrical power demand and increasing service quality as well as reducing pollution, the existing power grid infrastructure should be developed into Smart Grid (SG) that is flexible for interconnectivity with the distributed generation. However, integrating distributed generation to power system causes several technical issues especially system stability. To make the power grid become "smarter", particularly in terms of stability, Flexible AC Transmission System (FACTS) device especially Static VAR

Compensator (SVC) is used. This paper explores Smart Grid technologies and distributed generation systems. Furthermore, it discusses the impact of distributed generation on Smart Grid, particularly its system stability after installing distributed generation in the Smart Grid. This was done by examining the system stability during interconnection and faults on the system and validated with Dig-SILENT Power Factory Software V 13.2.

KEYWORDS

Smart Grid, Future Grid, Distributed Generation, Flexible AC Transmission System (FACTS), Static VAR System (SVS)

1. INTRODUCTION

The current electric power industry is undergoing major changes from being centralized generation into decentralized generation. The advances in technology have created rapid growth in the utilization of distributed generations which leads to energy market becoming more attractive and competitive. Moreover, due to electricity deregulation, environmental issues as well as government incentives, this technology has created interest in development further amongst industrial countries throughout the globe. In addition, these issues also results in the Smart Grid platform to end the traditional electric power industry which traditionally was vertically integrated and congested resulting in a higher energy costs.

Nowadays, most of aging and large remote power system stations with central dispatch suffers from disturbances due to lack of intelligent interoperability units. The system also becomes vulnerable when utility abnormalities are present, for example on protection or failures to control coordination and human operation errors. Therefore, there is a need to transform this model into Smart Grid that can enhance power quality and fully integrate with advanced grid elements such as intelligent sensing and digital metering.

Smart Grid is recognized as a new platform for future power industry. The rapid rise on this issue is also leading to the fast growth of distributed generation technology markets such as in fuel cells (FC), photovoltaic (PV), wind turbine (WT) and energy storage (ES). This trend will have profound impact on future electricity technology which allows Information and Communication Technologies (ICT) and advanced power electronic devices to be installed and embedded throughout the network. This is the challenge where current bulk generation and distributed generation will co-exist with higher power reliability and quality in the form of Smart Grid. To emphasize these, this paper will provide a fundamental understanding of distributed generation issues and framework of Smart Grid. Lastly, it finishes off by providing an analysis impact of DG on Smart Grid transient stability.

2. OVERVIEW OF DISTRIBUTED GENERATION, SMART GRID AND SVC TECHNOLOGIES

2.1. Distributed Generation

Distributed Generation (DG) technology incorporates wind turbines, micro turbines, photovoltaic systems, fuel cells, energy storage and synchronous generator applications to supply active power to distributed systems connected close to the consumers load. This concept is becoming a major player for Green House Gases (GHG) mitigation and power system reliability. Therefore, many developed countries such as Australia, are encouraging DG utilization for local power source and to avoid concentration of new power system transmission or distribution construction.

The increasing number of DG connection to the grid for the last decade proves that DG intrinsically offers various technical, economical and environmental advantages for customers, utilities and power operators. However, the presence of DG in the current existing power system has an impact on its operation and configuration. Moreover, since most of current aging power system is considered as passive system and with the insertion of DG, the network becoming active system where both of them can act as power sources.

Various studies in the past examined the impact of DG in power system and identified some vital aspects concerning their operation and connection. In reference [1] the analysis on change of power flow direction in corporation of DG integration is verified. In reference [2] the influence of DG on system reliability and stability during peak load is investigated whilst reference [3] focuses on system protection as the short-circuit fault level on the network tends to increase due to DG connection. Reference [4] investigates the voltage variation and protection by comparing the real system behavior with software simulation. The authors conclude that the level of transformer tap changer is increased with an increase on the penetration level of DG. Reference [5] has also reviews the impact of DG on distribution system (DS). At present DG provides much benefit in order to improve voltage quality, loss reduction and reliability. Reference [6] also shows that during interconnection to the distribution network, DG unit can be operated under island mode. It highlights the interconnected DG performance and shows this is greater during islanding condition, especially from security and quality of supply point of view. In addition, the study in [7] demonstrates the positive impact of DG on DS where during faults period, the rotor angle deviation and voltage drop are found to decrease. It means that the transient stability of the system improves with an increase in the penetration level of the DG.

However, overall these research scenario investigates the technical impact of DG on DS where almost all of them are coupled to the medium and low voltage levels. Moreover, none of the research scenario deals with the issues from Smart Grid perspectives. With the presence of Smart Grid concepts and rapid growth of DG and Smart Grid technologies, it is critical to evaluate the system performance precisely. Thus, the framework and concepts of Smart Grid can be applied appropriately and this avoid degradation of power quality and reliability.

2.2. Smart Grid [8]

Nowadays, the intelligent automation on electrical distribution network has been driving the development of future power system like Smart Grid, which might lead to the network model being changed and certainly this will change the electric network operations. Smart Grid is a relatively new concept for future power system that integrates electricity and communication on power system network, which supplies digital information on real time network operation for the operator and ultimately the consumer.

The main characteristics of Smart Grid are self-healing, empowering the customers, improving power quality and ability to accommodate various distributed generation. In addition, advanced control methods, digital sensing and metering, advanced grid devices such as FACTS (Flexible AC Transmission Systems) and SCADA (Supervision Control and Data Acquisition) system are few of the major technologies involved in the implementation of Smart Grid. FACTS is alternating current transmission system device used to enhance control ability and increase power transfer capability based on power electronic and other static controllers, meanwhile the SCADA system refers to the control and communication data system used for operation, monitoring and controlling power system grids. However, the standard of smart grid, how all elements are connected with each other and how the communication or energy flow works still remains the major concerns for enabling Smart Grid implementation.

The National Institute of Standard and Technology (NIST) reveals that there are seven important Smart Grid areas that need to be addressed to make the grid "Smarter" including; bulk generation, transmission, distribution, customer, operation, market domain and service provider [9]. These domains are integrated based on the conceptual model of Smart Grid as shown in **Figure 1**. As can be seen from the figure, it is obvious that Smart Grid is an integrated system which involves complex co-ordination strategy for implementation. In addition, Smart Grid technology broadens power knowledge and involves interdisciplinary research area such as: communication, automation, sensor and control. Furthermore, Smart Grid also requires advanced technology to make the grid as an ideal "Smart Grid". Nevertheless, due to the complexities of the system, it is unclear and even confusing to define the networks being "smart" if a few of its key characteristics are neglected. Instead, it is preferable to consider the term "Smart Grid" as the chance to enhance the power system performance and improve operational capabilities [8,10].

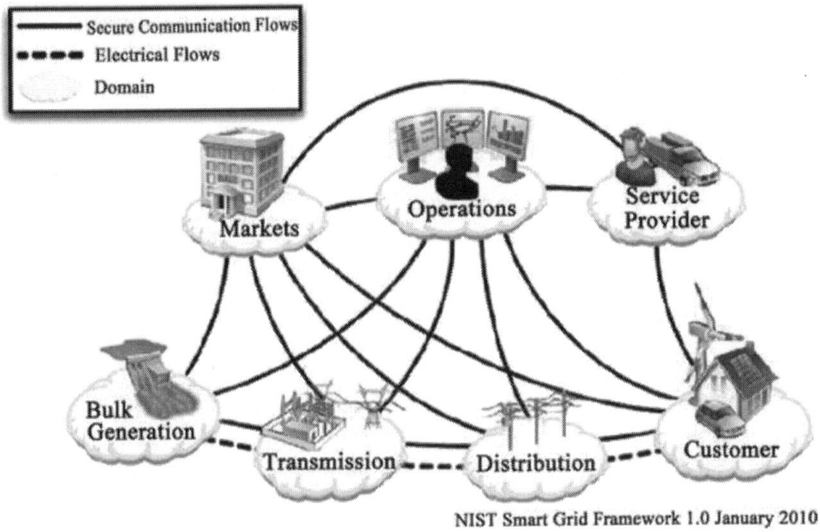

Figure 1. Smart Grid topology network [9].

2.3. Static VAR Compensator (SVC)

FACTS controller history initially began when Hingorani proposed the scheme of power compensation in electrical power system using power electronic applications [11]. Afterward, numerous researches were conducted on the application of FACTS using self commutated semiconductor and thyristor in transmission system. Originally, FACTS devices were developed for transmission systems which consisted of electrical conductors that have resistance, inductance and capacitance (R-L-C). The capacitive and inductive reactance in the transmission line generates and absorbs reactive power. The reactive power that flow along the line often causes further loss in the resistance of the conductor. For this reason FACTS device is required to enhance the power transfer capability and stability in the transmission system. However, for the last decade this concept has been extended for improving power quality on distribution systems operation as well [12]. The application shown in [12] is to control and maintain the system stability where in the distribution level most of the loads are non-linear or dynamic.

Flexible AC Transmission System is one of the advanced technologies used for power system compensation. The devices including Static VAR

Compensator (SVC), Static Synchronous Series Compensator (SSSC), Static Synchronous Compensator (STATCOM), Thyristor-Unified Power Flow Controller (UPFC), Thyristor Con-trolled Series Compensator (TCSC), Thyristor-Switched Series Capacitor (TSSC) and Thyristor Controlled Reactor (TCR) etc. These devices have many configurations, commonly it can be classified into series-connected controllers, shunt-connected controllers and a combination of them. The detail description of series and shunt-connected controller can be found in [11].

Static VAR Compensator (SVC) is "a shunt-connected static VAR generator or absorber whose output is adjusted to exchange capacitive or inductive current so as to maintain or control specific parameters of the electrical power system (typically bus voltage) [11]". SVC is based on thyristor-controlled and switched shunt component without gate turn-off capability. It is a variable impedance device using back to back connected thyristor valves to control the current flow through reactor. SVC as a control device offers fast response time and much faster than traditional mechanically switched reactors or capacitors. The configuration of SVC as shown in **Figure 2**consists of two main components and their combination:

- Thyristor-controlled and Thyristor-switched Reactor (TCR and TSR)
- Thyristor-switched Capacitor (TSC)

TCR and TSR constitute of a shunt-connected reactor controller using pair parallel back to back-connected thyristor. Using phase angle control, TCR generates an equilent and constant variable inductive reactive power from zero to maximum. Whereas, TSR is controlled without phase angle control which results in a step change in retance and provides fixed inductive admittance.

TSC has similar operational characteristic and composition as TSR. It consists of a back to back thyristor pair in series with capacitors. The TSC is not continuously conrolled because of transient phenomena at switch-on, however instead is switched on and off independently. Therefore, TSC cannot inject a reactive current with variable amplitude into the system. The transient phenomenon in TSC does not generate harmonics but if they appear, it isnot a serious problem [13].

3. Equal Area Criterion for Transient Stability Analysis

It is well known that large or small-scale integration of distributed generation may have significant impact on power system stability with respect to the rotor angle, voltage and frequency stability. Reactive power compensation and voltage control is fundamental to make the grid become smarter. Without this control, the presence of distributed generation may potentially cause system collapse. Therefore, a dynamic shunt reactive power compensator such as SVC is required to mitigate these issues since it can be used to enhance transient stability margin in power system. The potential effectiveness of SVC on transient stability enhancement can be examined through the following discussion.

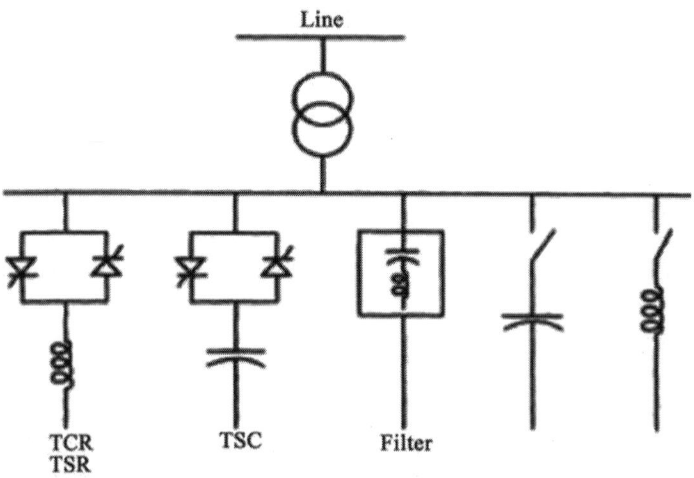

Figure 2. SVC configuration [11].

As depicted in **Figure 3**, assume that a single power generation machine with interconnecting lossless lines have a reactance (X). Denoting the terminal voltages of generator machine and the infinite bus by $V_1\angle\delta$ and $V_2\angle 0$, power angle (δ) and the transmitted power as P, then without the SVC installed in the line, the value of the power transfer can be expressed by,

$$P = \frac{V^2}{X}\sin\frac{\delta}{2} \quad (1)$$

To understand the advantages of SVC, assume that an SVC is installed at the mid-point of the interconnecting line. In that case, the reactance of the SVC between the machine and infinite bus is X/2 ohms. Subsequently, the value of transmitted power can be calculated as

$$\left[\frac{V^2}{(X/2)}\right]\sin\frac{\delta}{2} = 2\frac{V^2}{X}\sin\frac{\delta}{2} \quad (2)$$

From Equation (2) it can be seen that the value of transmitted power is doubled i.e., from V^2/X it goes to $2(V^2/X)$. Accordingly, the transient stability level is also increased. This fact can also be validated by the transient stability equal area criterion with SVC and without SVCas shown in **Figure 4**.

Consider that **Figure 4**(a) is system without SVC and **Figure 4**(b) is system with SVC. Assume that both sys-tems are subjected to the same fault for the same period of time. Before the fault, each system transmits power PM and is operating at angle δ1 and δc1. During the fault, the transmitted power decreases significantly while the mechanical input power remains constant (PM). As consequence, the generator accelerates from δ1 to δ2 and δc1 to δc2 at which the fault clears. After fault clearing, the transmitted power exceeds the mechanical input power and the generator starts to decelerate. Nevertheless, their angle increases due to the kinetic energy stored in the rotors. At δ3 and δc3 the maximum rotor angle are reached when the decelerating energies that represented by areas 'A2'and 'Ac2' become equal to the accelerating energies defined by areas 'A1' and 'Ac1'. The maximum of transient stability limits is achieved at δ3 and δccrit. The areas of A margin and Ac margin represent the transient stability margin of the system.

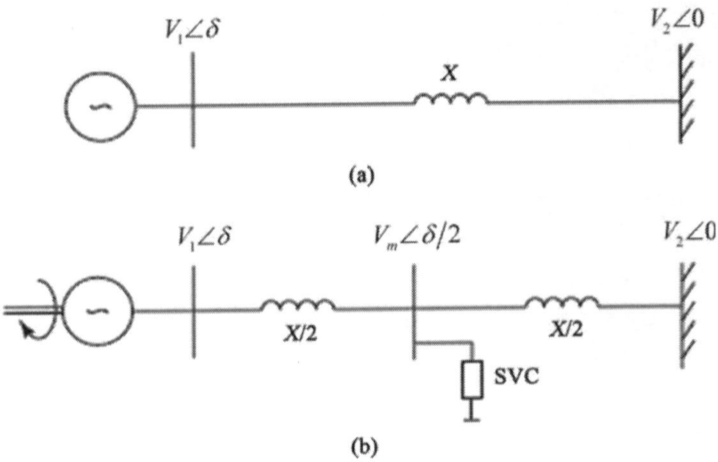

Figure 3. Simplified power systems (a) without SVC and (b) with SVC [14].

From above description, it is evident that SVC is able to enhance transient stability limit. In addition, it is obvious that the transient stability is determined by power (P) against power angle (δ).

4. EXPERIMENTAL SETUP

In order to make the power system become "Smarter", this research considers small and simple system configuration with easier control and design as shown in **Figure 5**. The one-line system network is derived from [6] which presents successful islanding operation of distributed generation into the Smart Grid environment. However, this research only focuses on islanding examination and does not consider analyzing the system stability. Therefore, it is necessary to developed further under Smart Grid concept by simplifying the system in respect of control where Flexible AC Transmission System (FACTS) devices particularly Static VAR Compensator (SVC) is used. The typical network intends to evaluate the impact of distributed generation on Smart Grid's steady state voltage profile and system stability or dynamic behavior, which is essential to establish an appropriate applied model.

Consider Smart Grid network consisting of a 132 kV; 50 Hz sub transmission system with maximum of short circuit levels 5000 MVA and 4000 MVA respectively. The grid feeds a 33 kV distribution system through a 132 kV/33 kV, Δ/Y_g transformers with rated power equal to $S_T = 90$ MVA, Short Circuit Voltage $(V_{cc}) = 13.18\%$. In the system, two type of distributed generation technologies are considered as dispersed sources; named DG I and DG II. Each of them has rated power of 28.1 MVA and are separately coupled into the 11 kV bus through 33/11 kV transformers. The grid load in Smart Grid zone is 1 MW while the maximum load of the DGs is 23 MW.

The SVC, which letter on this is called Static VAR

System (SVS), is installed in the grid to make the system become "smarter" particularly to maintain the system stability during the integration of distributed generation. **Figure 6** shows the model of SVS controller frame used in this simulation [16]. It is based on IEEE model with the objective for voltage and power reactive control in the system. The SVS is installed at bus 2 at the mid-point of the system. The basic data parameters are given in **Table 1**.

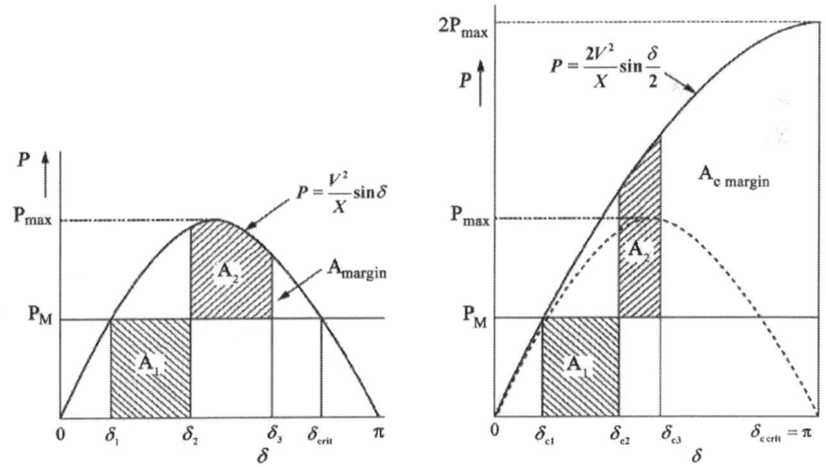

(a) (b)

Figure 4. Equal areas illustrating transient stability limit without SVC (a) and with SVC (b) [15].

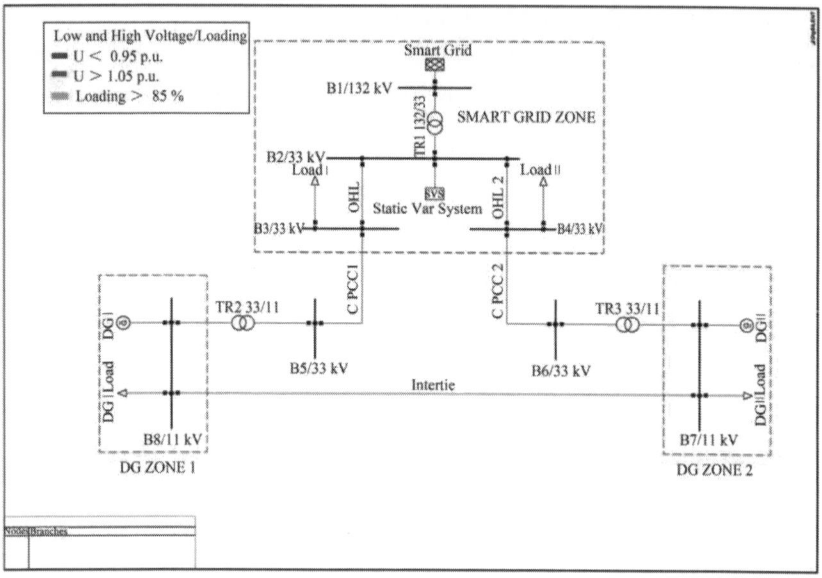

Figure 5. Smart Grid topology network.

Figure 6 shows that the measurement element (ElmSsc*) obtains active and reactive power, voltage, current and other signals from power system. The mea-surement system objective is to ensure that all the inputs are proportional for stable operation as required by the system. These signals then are transferred to SVS controller element (ElmSvc*) which has two distinct mode controls: voltage and reactive power control. The voltage control system aims to keep the voltage within acceptable limits, especially at Point of Common Coupling (PCC) since it is vital voltage regulation of the systems. To perform this, the voltage regulator processed all measured input variables and creates an output signal that is proportional to reactive-power compensation. The measured variables are compared with the voltage reference signal, and the error becomes an input variable forthe controller. The output of the controller creates a per-unit susceptance signal to reduce undesired signal error to zero in the steady state operation; afterward the signal is transferred to the gate-pulse generation. The output signal derived from voltage regulator is transferred to the gate-pulse generation which creates proper firing pulses for TCR-TSC of the SVC slot, subsequently the undesired susceptance signal is rectified and the desired susceptance output is available at SVC bus.

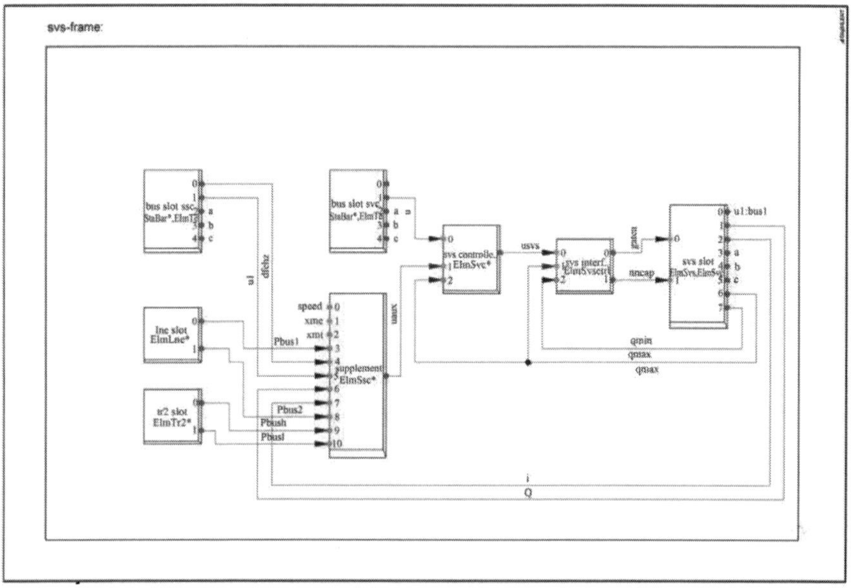

Figure 6. SVS control frame [16].

Table 1. Static VAR System parameters.

TCR	Q Reactance (>0)	3 MVAR
	TCR Max Limit	3 MVAR
TSC	Max Number Capacitors	2
	Q per Capacitor Unit (<0)	–2 MVAR

5. CASE STUDIES

This section provides general overview of the case scenarios for the research and its assumption. In order to illustrate the impact types of distributed generation on Smart Grid, steady state voltage analysis and transient stability analysis have been performed. The considered scenarios are as follow:

Case 1: Steady state voltage profiles have been evaluated using a load flow Newton-Raphson Method to check the voltage pattern and possibly overloading.

Case 2: Transient stability analysis was performed particularly when DG I and DG II are coupled into Smart Grid. The parameters being measured are frequency, voltage and power angle.

Case 3: Transient stability analysis was performed particularly when fault occurs in the system. The parameters being measured are frequency, voltage and power angle.

Before observing the impact of DG on Smart Grid the following procedure was adapted to address the 3 cases presented above. In all cases, steady state profiles have been evaluated using a load flow Newton-Raphson procedure. The distributed generation is modeled in PQ and PV node. The system is assumed to be in very high demand and under balanced condition. The dynamic behavior of load is not considered.

Initially, during operation the voltage and frequency of the DG I is controlled by Smart Grid, thus the DG I at 11 kV bus is operated in PQ mode. Meanwhile, DG II operates independently in PV mode. For safety reason, normally DG should not operate in PV mode when it is connected to network. However, in this case study the operation mode of DG II is kept in PV mode when it is connected to the grid. This approach is intended to evaluate the reliability of connecting DGs to Smart Grid when DGs modeled as PV node. In all case studies, DG II is connected to the grid at t = 4 seconds by closing the Point of Common Coupling (PCC) in bus 7 and the intertie line between GT I and GT II. In addition, to avoid system collapse due to power reactive injection or absorption from distributed generation the SVS is operated in reactive power control mode (Q mode).

6. EXPERIMENTAL RESULT AND ANALYSIS

The following section presents the result and analysis done for the simulations. In all simulated case studies, the load flow calculation and initial system conditions have been performed successfully. At this stage, preliminary simulation shows that there was no error found in the system topology such as over voltage or under voltage and over load. The default information in this regard is shown in **Figure 7" target="_self"> Figure 7**.

6.1. Smart Grid Steady State Voltage Analysis (Case 1)

One of the main benefits from integration of distributed generation in electricity networks is an improvement of the steady state voltage profile of the system. However, voltage violation and fluctuation during the connection of distributed generation could happen and considerably influence the network transient stability. Therefore, the Power System Operators (PSO) must verify that in the worst situation, the networks steady state voltage variation is no greater than 5%. This section addresses the impact of distributed generation on Smart Grid steady state voltage.

Regarding Smart Grid steady state voltage profile before employing DG I and DG II, it is assumed that in all case studies the entire system bus voltage remain constant at 1 p.u. The considered maximum and minimum voltage variation is ±5% from nominal voltage. Thus, the results of steady state voltage profile obtained for all buses after employing DG I and DG II is presented in **Table 2**.

It can be seen from **Table 2** that a voltage deviation in Smart Grid still occurs due to imbalance in reactive power generated by the DGs during the integration. However, large violations of voltages are not evident on all buses. The overall Smart Grid voltage remains within the allowable range between 0.95 p.u and 1.05 p.u, which is 0.996 p.u. From the result presented in **Table 2** it appears that the bus 1 (SG zone) is not sensitive when DG I and DG II units are utilized in Smart Grid.

6.2. Smart Grid Transient Stability Analysis

The Smart Grid response following integration of distributed generation to the network steady state profile has been examined. It demonstrates that the Smart Grid steady state voltages profile is within acceptable limit between 0.95 p.u and 1.05 p.u. Nevertheless, this condition cannot guarantee the system remains stable if a fault occurs on the system. Therefore it is critical to examine the Smart Grid transient stability. The transient stability analysis investigates the system responses after distributed generation is integrated to Smart Grid and following fault occurrence. The considered indicator for stability study is determined by analyzing Smart Grid voltage magnitude, frequency and power angle.

6.2.1. Grid Connected Mode Stability Analysis (Case 2)

This section discusses the Smart Grid stability (Case 2) during integration of distributed generation or grid connected mode stability analysis. **Figure 8** shows that by connecting DG II the voltage stability of Smart Grid at bus 132 kV changes slightly. However, the deviation remains in acceptable range and clearly shows that the voltage takes 5 seconds to stabilize after connecting to DG II. The presence of voltage instability in this case results from progressive rise in operating slip or speed of distributed generation to restore the power and synchronizing the system as illustrated in **Figure 9**. As a consequence, due to high rotor slip the reactive power demand by DG II is greatly increased and might cause further voltage instability if reactive power controller is not used. Therefore, during grid connected mode, the SVS is operated in Q mode to provide and regulate the reactive power.

Following voltage imbalance, further studies were carried out to analyze the frequency stability of Smart Grid. As illustrated in **Figure 10**, it can be seen that the frequency level increase slightly and occurs at switching point where DG units are integrated to Smart Grid.

```
DIgSI/info - Element 'Grid\Smart Grid.ElmXnet' is local reference in separated area 'Grid\B1\132 kV.StaBar'
DIgSI/info - Calculating loadflow...
DIgSI/info - ----------------------------------------------------------------------------------
DIgSI/info - Start Newton-Raphson Algorithm...
DIgSI/info - load flow iteration: 1
DIgSI/info - load flow iteration: 2
DIgSI/info - Newton-Raphson converged with 2 iterations.
DIgSI/info - Loadflow calculation successful.
DIgSI/info - Element 'Grid\Smart Grid.ElmXnet' is local reference in separated area 'Grid\B1\132 kV.StaBar'
DIgSI/info - Element 'Grid\Smart Grid.ElmXnet' is reference in 50.0 Hz-system
DIgSI/info (t=000:000 ms) - Initial conditions calculated.
```

Figure 7. Information of initial load flow and system conditions.

Table 2. Overall Smart Grid voltage profile after employing distributed generation.

Bus No	Before (p.u)	After (p.u)
1	1.0	1.0
2	1.0	0.966
3	1.0	0.966
4	1.0	0.966
5	1.0	0.966
6	1.0	0.966
7	1.0	0.966
8	1.0	0.966
9	1.0	0.966

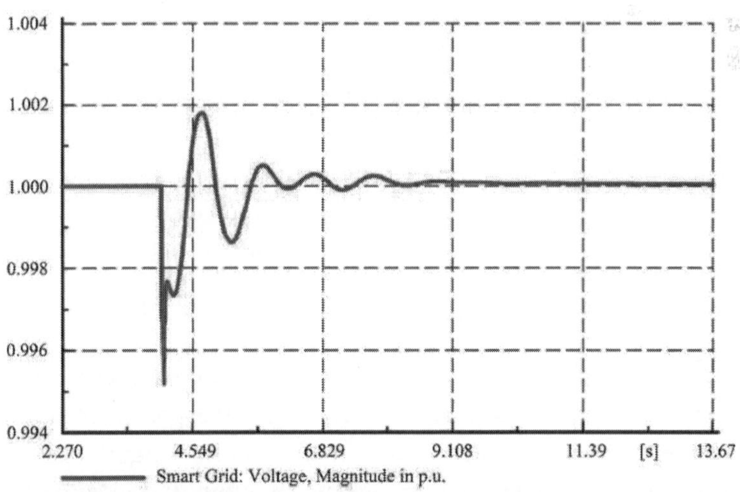

Figure 8. Smart Grid voltage magnitude in p.u during DGs integration.

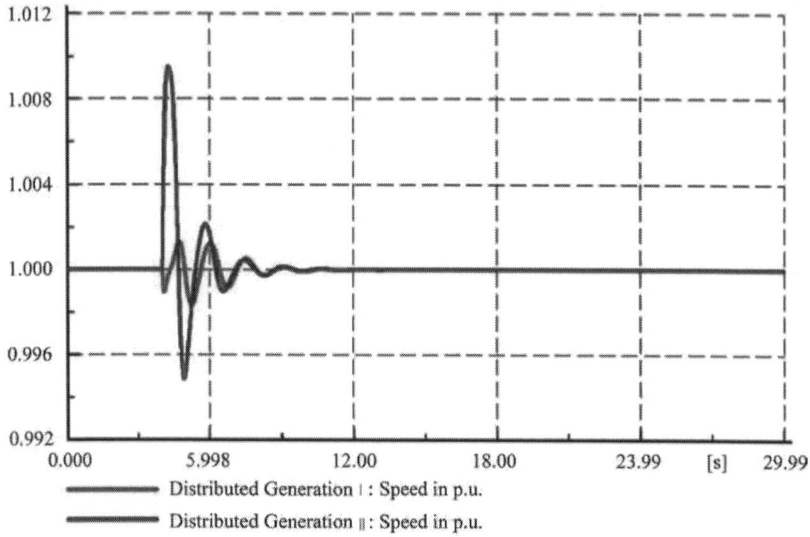

Figure 9. Distributed Generation speed in p.u during integration to Smart Grid.

However, the network succeeded to maintain new steady frequency after integrating with DGs. In this case, the frequency rises from 50 Hz to 50.03 Hz when the PCC was closed at t = 4 seconds.**Figure 11** shows the range of frequency deviation, which is only 0.04 Hz and is no more than ±2 Hz from nominal 50 Hz. Therefore, this frequency variation is acceptable and most of end user can bear this change. The Grid Code requires that the system frequency must be controlled within the limits of 47.5 - 52 Hz in case the system frequency rises above the limit [17]. However, larger violation can result in network collapse which leads to blackout.

On the other hand, another indicator to asses Smart Grid transient stability during grid connected mode is the voltage angle stability. As can be seen from **Figure 12**, the integration of DGs unit causes change of voltage angle which forces the network to operate at a new operating point. In this case, the maximum deviation was at t = 4 seconds between 89.60 and 90.22 degrees. The amount of reactive power absorption or injection by DGs during interconnection determines a significant respond of power angle, voltage and frequency stability. If the system lacks sufficient reactive power, it could cause an under voltage. However, it can be seen that the network maintains the voltage angle back to its original structure after 5 seconds.

6.2.2. Fault Stability Analysis (Case 3)

With the same network configuration and SVS connected, further Smart Grid stability is examined after the integration of DGs. The 200ms self-clearing 3 phase fault short circuit was applied at 15 seconds in the middle of transmission line connecting between Buses 2 and 3. The Smart Grid response during this fault is depicted in **Figure 13**. It can be seen that during the fault, the AC voltage plunges from 1.0 p.u to 0.988 p.u and it causes the electric torque to drop. Accordingly, the DG rotor speeds up and the reactive power absorption is greatly increased because of high rotor slip. However, due to fast dynamic power reactive support from the SVS, the Smart Grid voltage recovers within 2 seconds after the clearance of the fault and the system goes back to stable.

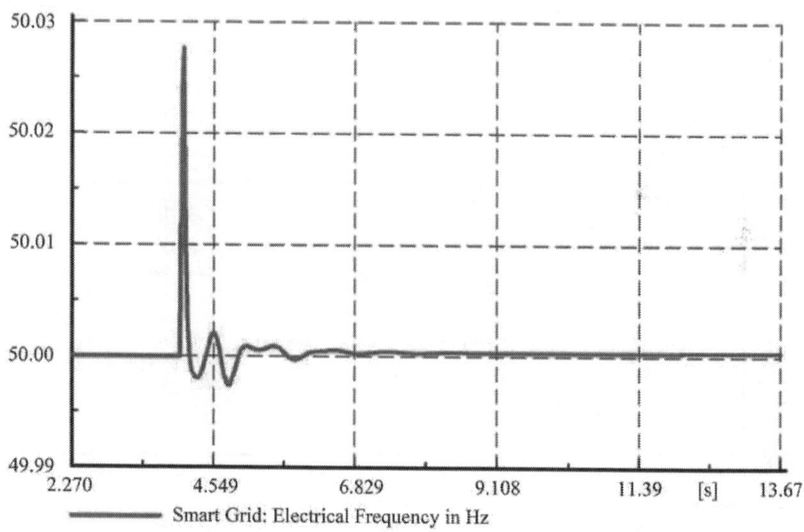

Figure 10. Smart Grid frequency during integration of DGs in Hz.

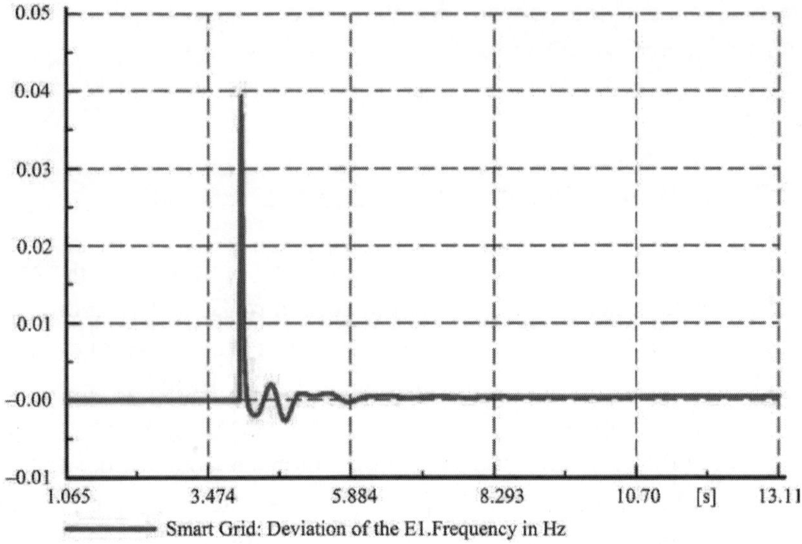

Figure 11. Smart Grid frequency total deviation during integration of DGs in Hz.

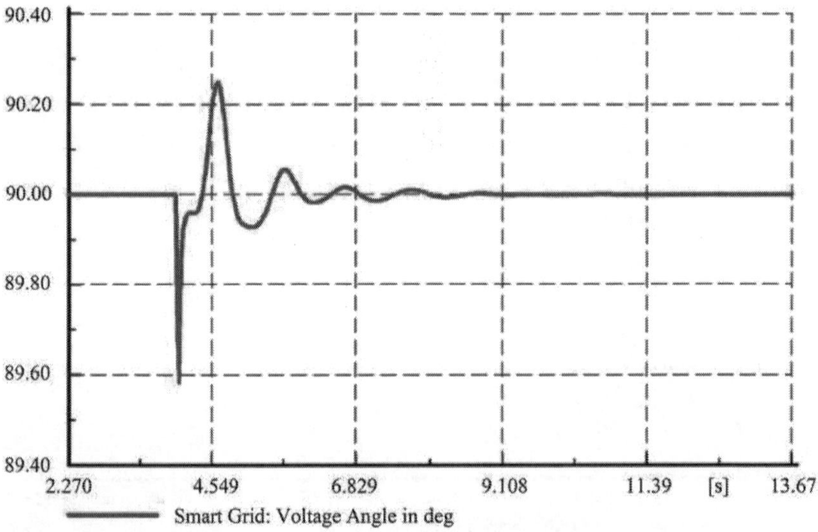

Figure 12. Smart Grid voltage angle during integration of DGs in degree.

For the Smart Grid frequency stability response following the fault, **Figure 14** indicates that during the fault, the system frequency plunges from 50 Hz to 49.997 Hz. Unfortunately, the post-fault frequency is slightly less than its nominal value between 49.998 Hz and 49.999 Hz. This phenomenon occurs as consequence of imbalance between the electromagnetic and the mechanical torque of DGs after the fault. However, it tends not to be of significant concern. Moreover, the rotor speed is gradually reduced and their equilibrium also restored gradually after the SVS has provide sufficient reactive power support. Thus, the post fault frequency level remains stable above 49.998 Hz.

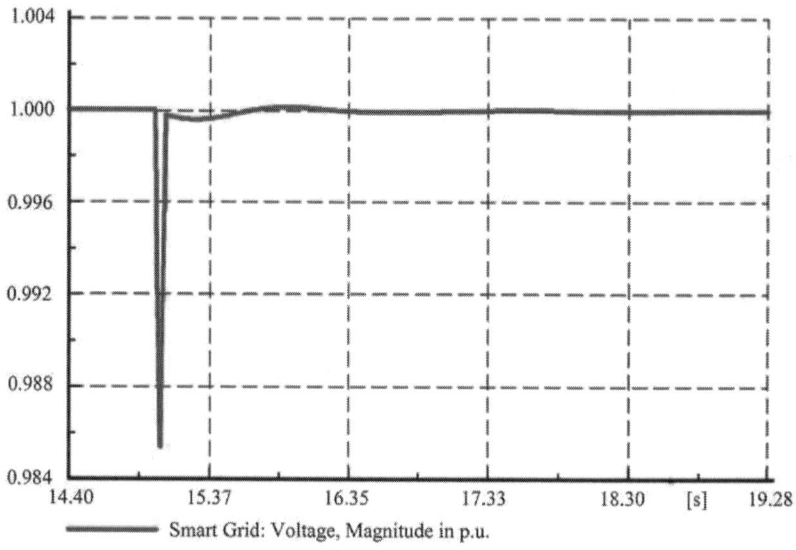

Figure 13. Smart Grid voltage responds during 3 phase fault in p.u.

On the other hand, as consequence of three phase short circuit fault, the Smart Grid voltage angle dips approaching to 89.30 degrees as depicted in **Figure 15**. This incidence leads to voltage drop as illustrated in **Figure 13**. In view of this, the change of the voltage angle will always correspond with the change of voltage and frequency limits. However, the use of SVC helps the system to maintain synchronism with the DGs during and post fault. As shown in **Figure 15**, thc voltage angle regains to 90 degrees within 2 seconds. Thus new steady state value between DG and Smart Grid reaches.

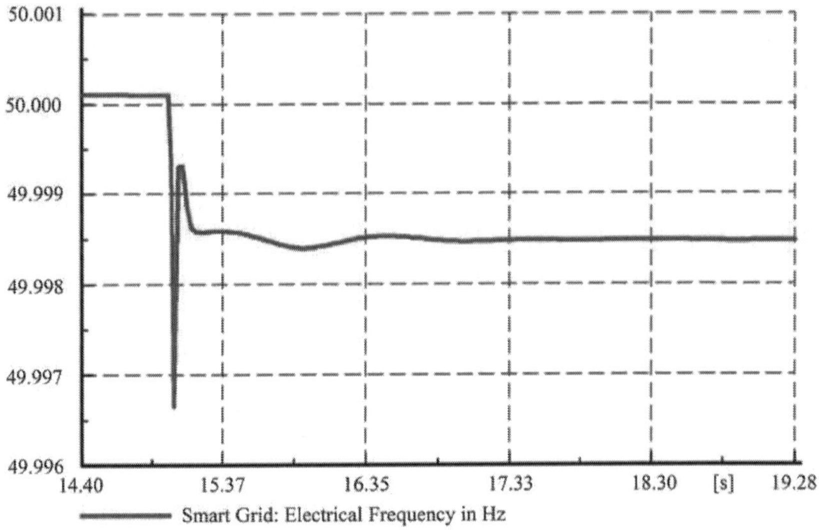

Figure 14. Smart Grid frequency responds during 3 phase fault in Hz.

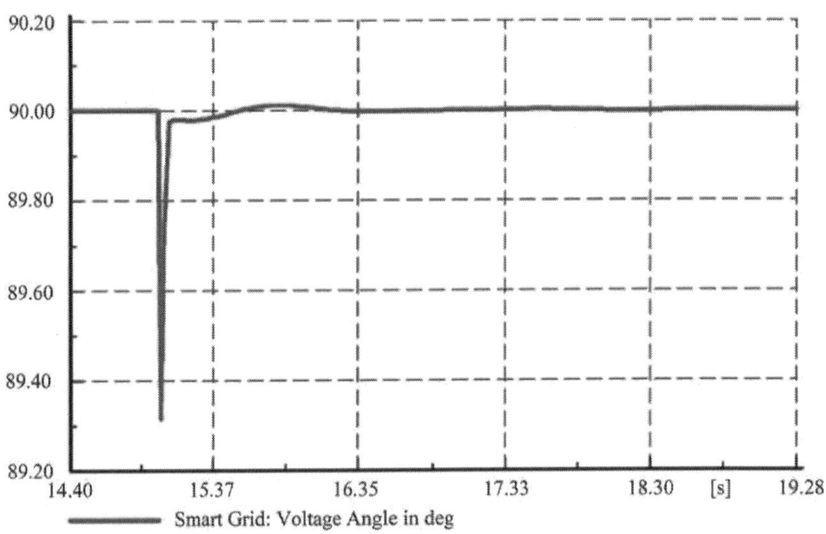

Figure 15. Smart Grid voltage angle during 3 phase fault in p.u.

7. CONCLUSIONS

This paper addresses the impact of distributed generation on Smart Grid's transient stability analysis. The steady state voltage during the integration of DG and transient stability of Smart Grid following self-clearing 3 phase fault have been examined.

Regarding steady state voltage behaviors, it is found that the integration of DG does not modify the performance of Smart Grid. It shows that the steady state voltage deviation is still within limits. Transient stability analysis however, shows that the utilization of SVC is able to maintain the voltage, frequency and voltage angle and remains in stability margin following the DG connection and fault. The simulation result clearly indicates that reactive power support is necessary when utilizing DG. This fact is confirmed through transient stability equal area criterion in Section 3 that the existence of SVC helps significantly the transient stability of the system. With proper control and installation of SVC in the system, it can provide voltage and power reactive support and with the increase in transmission capability transient stability margin increases. In addition, the feasibility of connecting DG in PV mode was also analyzed. It can be seen that the DG II is successfully integrated in Smart Grid while its operation is in PV mode; therefore the DG II is not switched in PQ mode operation during integration to the grid.

Finally, the results of this research shows that in particular term of stability and time recovery, a favorable comparison to reference [6] and [18], which takes more than 8 seconds to settle after islanding. Due to its capability for successful DG integration and a rapid recovery after fault, it is suitable to be examined as Smart Grid. However, further investigation for future work is still necessary; the scope of the research includes harmonic analysis, different FACTS controller and level of DG penetration.

ACKNOWLEDGEMENTS

The first author would like to acknowledge the Australian Development Scholarship (ADS) for providing the support essential to pursue his Master by Research at Victoria University, Melbourne, Australia.

REFERENCES

1 T. Khoan and M. Vaziri, "Effects of Dispersed Generation (DG) on Distribution Systems," IEEE in Power Engineering Society General Meeting, Sacramento, 16 June 2005, Vol. 3, pp. 2173-2178.

2 R. A. Prata, "Impact of Distributed Generation Connection with Distribution Grids—Two Case-Studies," IEEE in Power Engineering Society General Meeting, Montreal, Quebec, Vol. 19, 18-22 June 2006, p. 8.

3 C.-I. Chai, W.-J. Lee, P. Fuangfoo, M. Williams and J. R. Liao, "System Impact Study for the Interconnection of Wind Generation and Utility System," IEEE Transactions on Industry Applications, Vol. 41, No. 1, 2005, pp. 163-168.

4 F. Bastiao, P. Cruz and R. Fiteiro, "Impact of Distributed Generation on Distribution Networks," Proceedings of the 5th International Conference on European Electricity Market, Lisboa, 28-30 May 2008, pp. 1-6.

5 P. P. Barker and R. W. De Mello, "Determining the Impact of Distributed Generation on Power Systems. I. Radial Distribution Systems," Proceedings of IEEE Power Engineering Society Summer Meeting, Seattle, Vol. 3, 16-20 July 2000, pp. 1645-1656.

6 S. P. Chowdhury, S. Chowdhury, C. F. Ten and P. A. Crossley, "Operation and Control of DG Based Power Island in Smart Grid Environment," Proceedings of the 20th International Conference and Exhibition on Electricity Distribution - Part 1, Prague, 8-11 June 2009, pp. 1-5.

7 A. A. Kumar, A. R. Massannagari, A. K. Srivastava and N. N. Schulz, "Impact of Biomass Based Distributed Generation on Electrical Grid," Proceedings of Clean Technology Conference, Boston, 1-5 June 2008, pp. 222-225.

8 N. A. Hidayatullah, Z. J. Paracha and A. Kalam, "Impacts of Distributed Generation on Smart Grid," Proceedings of the International Conference of Electrical Energy and Industrial Electronic System, Malaysia, 7-8 December 2009, pp. 218-221.

9 NIST Special Publication 1108, "NIST Framework and Roadmap for Smart Grid Interoperability Standard, Realease 1.0," Office of the National Coordinator for Smart Grid Interoperability, National Institute of Standart and Technology, U.S Department of Commerce, January 2010.

10 C. W. Potter , A. Archambault and K. Westrick, "Building a Smarter Smart Grid through better Renewable Energy Information," Proceedings of IEEE/PES Power Systems Conference and Exposition, Seattle, Washington D. C., 15-18 March 2009, pp. 1-5.

11 N. G. Hingorani and L. Gyugyi, "Understanding FACTS: Concepts and Technology of Flexible AC Transmission Systems," IEEE Press, New York, 2000, p. 452.

12 K. R. Padiyar, "Facts Controllers in Power Transmission and Distribution," New Age International Ltd. Publisher, New Delhi, 2009, p. 532.

13 E. H. Watanabe, et al., "Flexible AC Transmission Systems," Power Electronics Handbook, Elsevier Inc, 2007.

14 R. M. Marthur and R. K. Varma, "Thyristor-Based FACTS Controllers for Electrical Transmission Systems," IEEE Press and Wiley, New York, 2002, p. 493.

15 R. S. Vedam and M. S. Sarma, "Power Quality VAR Compensation in Power Systems," CRC Press, Boca Raton, 2009, p. 268.

16 Dig-SILENT Power Factory Software 13.2.

17 National Grid Electricity Transmission plc, "The Grid Code," Issue 4, Revision 3, London, 2010.

18 I. Vokony and A. Dan, "Examination of Smart Grids in Island Operation," IEEE Bucharest Power Tech Conference, Bucharest, 2 June-28 July 2009, p. 1-7.

CHAPTER 6

Optimal Scheduling for the Complementary Energy Storage System Operation Based on Smart Metering Data in the DC Distribution System

Bokyung Ko [1], Nugroho Prananto Utomo [1], Gilsoo Jang [1,*], Jaehan Kim [2] and Jintae Cho [2]

[1] School of Electrical Engineering, Korea University, Seoul 136-713, Korea

[2] Korea Electric Power Research Institute (KEPRI), Daejeon 305-760, Korea

ABSTRACT

The increasing penetration of distributed generation (DG) sources in low-voltage grid feeders causes problems concerning voltage regulation. The penetration of DG sources such as photovoltaics (PVs) in the distribution system can significantly impact the power flow and voltage conditions on the customer side. As the DG sources are more commonly connected to low-voltage distribution systems, voltage fluctuations in the distribution system are experienced because of the DG fluctuation and uncertainty. Therefore, the penetration of DGs in distribution systems is often limited by the required operating voltage ranges. By using an energy storage system (ESS), voltage fluctuation can be compensated for, thus satisfying the voltage regulation requirements. This paper presents an ESS scheduling algorithm based on the power injection data obtained from a smart metering system. The proposed ESS scheduling algorithm is designed for use within a direct current (DC) distribution grid, which comprises customers, each with a PV and an ESS system. The purpose of this ESS scheduling algorithm is to optimize the ESS scheduling by considering the complementary operation among all the ESSs.

KEYWORDS

energy storage system (ESS); ESS scheduling; distributed generation (DG); power flow; DC power flow; decoupled power flow

1. INTRODUCTION

Various renewable energy generation methods have been developed to reduce greenhouse gas emissions. As renewable energy generation, which is susceptible to changes in the weather and environmental conditions, is expected to be increasingly connected to distribution systems in the future, voltage regulation problems in the distribution systems are receiving increased attention. High penetration of renewable energy generation in a distribution system can cause voltage problems because of the fluctuation in the active power injection [1].

In the Korea Electric Power Corporation (KEPCO) system, the photovoltaic (PV)-concentrated areas in low-voltage distribution systems, such as the Cheon-an distribution system, generally experience overvoltage problems. This suburban area has 36 houses, each with 2.5 kW of PV panels on one phase (A phase). During high insolation periods, excessive power from the PVs flows along each branch; this reverse power flow causes a voltage rise. As shown in Figure 1, the A phase voltage in this KEPCO distribution system experiences a higher voltage when compared with the other phases during periods of high PV generation, therefore, the PV penetration must be limited in order to maintain a reliable distribution system voltage.

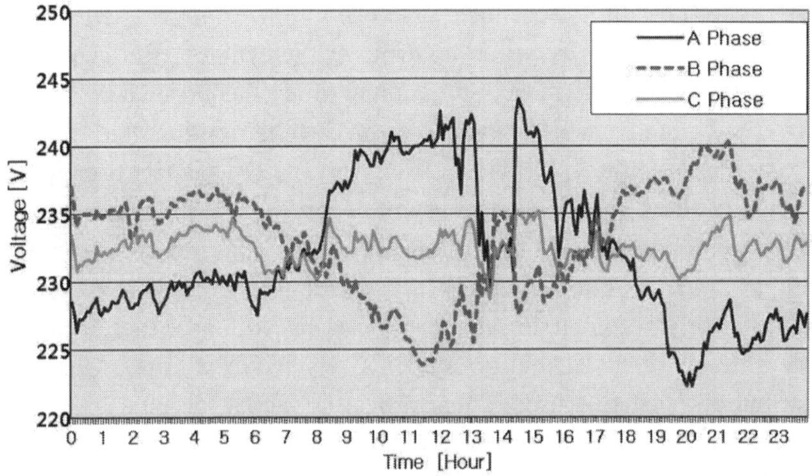

Figure 1. Voltage profile, including photovoltaic (PV) generation, within KEPCO's low-voltage distribution system during a day (A phase connected with many PVs).

This paper presents a scheduling algorithm for an energy storage system (ESS) system in a direct current (DC) grid that is installed to regulate the distribution system voltage for optimal operation despite any fluctuations in the PV generation [2]. This algorithm is based on the power injection information obtained from smart metering devices. The smart metering devices measure the power injection at each point of common coupling (PCC) and transmit the information to a data concentrator. With the presented system, the optimal ESS charging schedule can be determined by satisfying the voltage regulation requirements. The main objective of this paper is to propose an ESS control algorithm by considering complementary operation among all the ESSs, and not only the individual ESS control.

2. IMPACT OF ESS SCHEDULING ON THE DISTRIBUTION SYSTEM VOLTAGE

2.1. DC Distribution System

The distribution network at the Cheon-an site is a candidate for DC distribution implementation because of its high PV penetration. The DC system, which requires only voltage control, can benefit from system controllability because delivering power in the alternating current (AC) system requires control on the voltage, frequency and phase angle. Furthermore, the upper-side feeder-header AC/DC converter, which enables regulation of the distribution line voltage, has another benefit. In terms of efficiency, the DC distribution system is considered as a prospective system according to the increase in the DC-based loads. The DC distribution system has many advantages such as feasible connection of the distributed energy sources and reduction of the conversion losses for DC-based loads; therefore, it is very suitable for distribution systems, especially for the Cheon-an distribution system case. Therefore, this paper assumes that the distribution system is a DC distribution environment. As shown in Figure 2, the DC distribution system is composed of power electronic converters and a DC link. The ESS systems are applied at every PCC to compensate for the drawbacks of the DG systems, which affect the distribution system voltage.

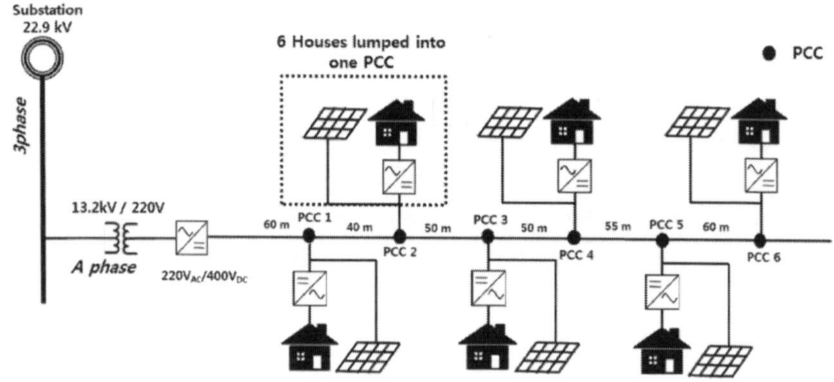

Figure 2. Configuration of the DC distribution system.

2.1.1. Efficiency Analysis for the DC Distribution System

In order to verify the efficiency of the DC distribution system, the efficiency comparison analysis between the AC and the DC distribution configuration, which is composed of residential loads, is performed. As summarized in Table 1, the composition of a residential load is assumed to be divided into three types: purely DC, DC conversion requiring AC, and purely AC loads. The composition ratio of each type of load is assumed to be 30%, 40% and 30%, respectively. In addition, the conversion step of each type of load is different according to the distribution system configuration, whether it is AC or DC. With regard to the converter efficiency, two types of efficiency are considered. One is the efficiency regarding the type of converter. As shown in Figure 3, the higher the capacity of a converter, the higher the efficiency is, and the AC/DC converter has the lower efficiency when compared with the DC/DC and DC/AC converters. The other efficiency is based on the loading conditions of a converter, as shown in Figure 4. By considering these factors to calculate the distribution system efficiency, it is verified that the DC distribution system can have benefits in terms of system efficiency.

Table 1. Power conversion efficiency for each type of distribution configuration.

Load type	Composition ratio	Conversion step		Efficiency	
		AC system	DC system	AC system	DC system
DC	30%	AC/DC and DC/DC	DC/DC	67.49%	89.38%
DC converted to AC	40%	AC/DC and DC/AC	DC/AC	69.60%	90.77%
AC	30%	-	DC/AC	-	90.54%
PV		DC/AC	DC/DC	90.93%	89.69%
AC/DC feeder-header converter		-	AC/DC	-	88.54%
Total distribution system efficiency				81.51%	83.63%

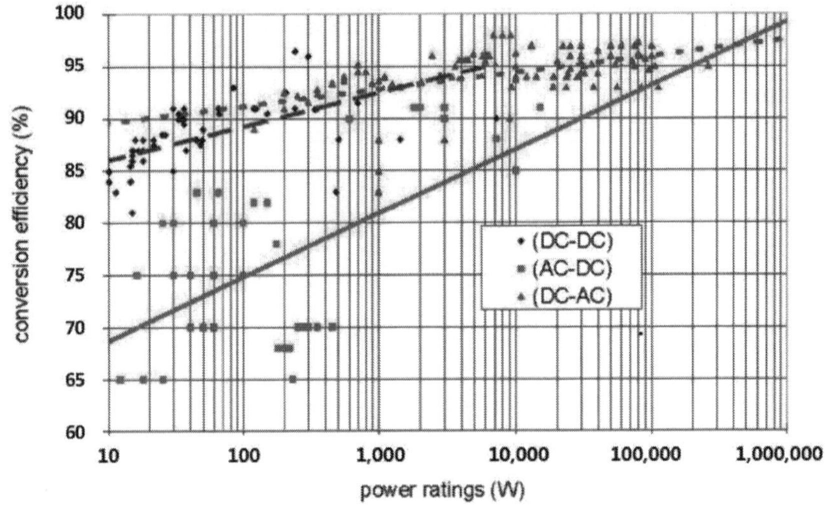

Figure 3. Converter efficiency for each converter type.

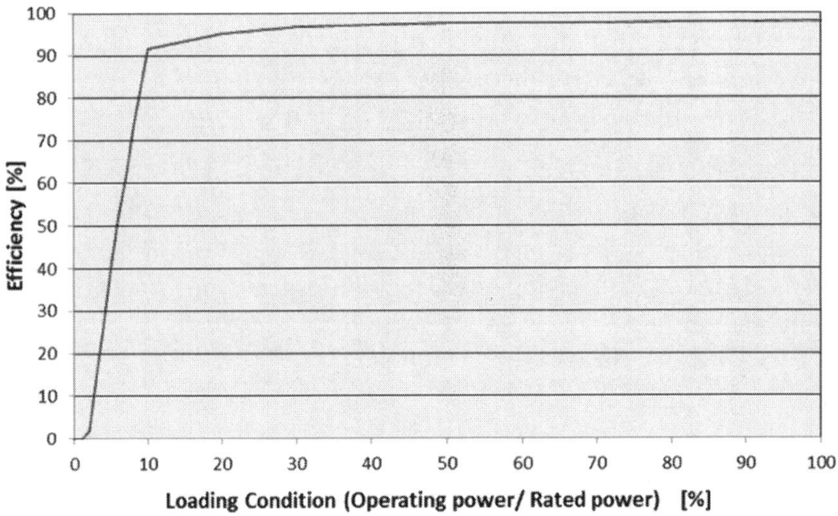

Figure 4. Converter efficiency according to the loading condition.

2.2. Impact of Distributed Generation on the Distribution System Voltage

The increasing diffusion of DGs such as PVs introduces voltage regulation problems in distribution systems. When the penetration of DGs is high in a distribution system, the system has the possibility of the distribution line voltage increasing and exceeding the allowable voltage limit because of the reverse power flow. DG based on renewable energy sources is susceptible to changes in the weather and environmental conditions, therefore, the amount of power generated depends strongly on those conditions. For this reason, the amount of voltage fluctuation is difficult to predict.

In the KEPCO system, the PV-concentrated areas in the low-voltage distribution systems, such as the Cheon-an distribution system, generally experience high-voltage phenomena. During periods of high insolation, excessive power from the PV sources flows along each branch; this reverse power flow causes a rise in voltage. The voltage at the A phase, which has many PVs in the distribution line, experiences a higher voltage when compared with the other phases during periods of high PV generation. As the KEPCO operating voltage range is restricted, the PV penetration is limited by the voltage regulation problems.

The extent of voltage fluctuation depends on the power system condition such as the distribution line parameters, distribution pattern of the DG, and output generation of the DG devices. Equation (1) shows the interrelation between the voltage fluctuation and the power inflow in a radial distribution power system. On the basis of Equation (1), when the power injection is concentrated on the lower side of distribution line, the extent of voltage fluctuation is more severe [3]:

$$\begin{aligned}\Delta V_n &= i_1 Z_1 + i_2 \sum_{k=1}^{2} Z_k + \cdots + i_n \sum_{k=1}^{n} Z_n \\ &= Z_1 \sum_{k=1}^{n} \frac{P_k}{V} + Z_2 \sum_{k=2}^{n} \frac{P_k}{V} + \cdots + Z_n \frac{P_n}{V} \\ &= \frac{P_1}{V} Z_1 + \frac{P_2}{V} \sum_{k=1}^{2} Z_k + \cdots + \frac{P_n}{V} \sum_{k=1}^{n} Z_n\end{aligned} \tag{1}$$

2.3. Effect of the Energy Storage System on Voltage Regulation

High generation within DGs and low load conditions lead to inversion of the power flow, which results in voltage rises. In order to solve this issue, this paper suggests the use of an ESS system for voltage regulation. The simulated system data are based on the Cheon-an distribution system, which frequently experiences voltage rise problems. The ESS can be used to store part of the energy produced during peak-generation periods by the PV and re-dispatch it during the evening or overnight.

This paper presents an algorithm for the ESS scheduling of power required for voltage support. This scheduling algorithm is based on the power injection data measured at the PCCs. These data are measured by a smart metering system and transmitted to a data concentrator, which estimates the optimal ESS scheduling condition while performing distribution line voltage regulation. By using the accurate system data from a smart meter, the current condition of the distribution system can be more precisely estimated. The main objective of the presented ESS scheduling algorithm is to control all the ESSs complementally, and not individually. Therefore, the optimized capacity of each ESS for voltage regulation can be derived [4,5].

3. OPTIMIZED ESS SCHEDULING TO COMPENSATE THE FLUCTUATION IN THE DISTRIBUTION SYSTEM VOLTAGE

3.1. Power Flow Calculation for the DC Distribution System

As the DGs are more commonly connected to low-voltage distribution systems, voltage stability has become a critical issue for power system operation. Voltage regulation can be disrupted by the fluctuations in the power inflow from the DGs. The assessment of voltage supported by ESS to ensure adequate distribution system voltage profile must be conducted with a power system analysis. Most power system analyses are based on power flow calculations. Therefore, to estimate a complementary charging schedule for the ESSs required for voltage regulation, system analysis based on the voltage and power loading conditions is required in advance. This section presents a methodology for the DC distribution system power flow estimation in order to acquire accurate information on the distribution system condition. The DC power flow analysis is a method to determine the optimal ESS operation schedule for all the ESSs in the distribution system to operate complementally for voltage regulation [6,7].

As shown in Figure 5, the power flow calculation is based on Kirchhoff's circuit law in electric power networks and comprises a set of nonlinear algebraic equations representing real, reactive power injection flow at each node. The nonlinear equations can be solved with an iterative algorithm such as the Newton-Raphson method. Equation (2) represents the AC power flow calculation, which is formulated with four variables at each bus: voltage magnitude, angle, and real and reactive power injection. The power flow calculation equation is a complex equation; therefore, it is necessary to separate the real and imaginary parts in order to apply the Newton-Raphson method:

$$
\begin{aligned}
P_k &= V_k \sum_{i=1}^{N} V_i [G_{ki} \cos(\theta_k - \theta_i) + B_{ki} \sin(\theta_k - \theta_i)] \\
Q_k &= V_k \sum_{i=1}^{N} V_i [G_{ki} \sin(\theta_k - \theta_i) - B_{ki} \cos(\theta_k - \theta_i)]
\end{aligned}
\qquad (2)
$$

Like the AC power flow calculation, the power flow in DC system can be expressed as Equation (3). Only two variables, voltage and real power injection data, are required to calculate the DC power flow because the DC system has no components related to reactive power and inductance. Therefore, the reactive power term (Q_k) and inductance term (B_{ki}) are excluded from Equation (2). The voltage angle difference ($\theta_k - \theta_i$) is zero:

$$P_k = V_k \sum_{i=1}^{N} V_i [G_{ki}] \tag{3}$$

$$\begin{bmatrix} I_1 \\ I_2 \\ I_3 \\ \vdots \end{bmatrix} = \begin{bmatrix} G_{11} & G_{12} & G_{13} & \cdots \\ G_{21} & G_{22} & G_{23} & \cdots \\ G_{31} & G_{32} & G_{33} & \cdots \\ \vdots & \vdots & \vdots & \ddots \end{bmatrix} \begin{bmatrix} V_1 \\ V_2 \\ V_3 \\ \vdots \end{bmatrix} \tag{4}$$

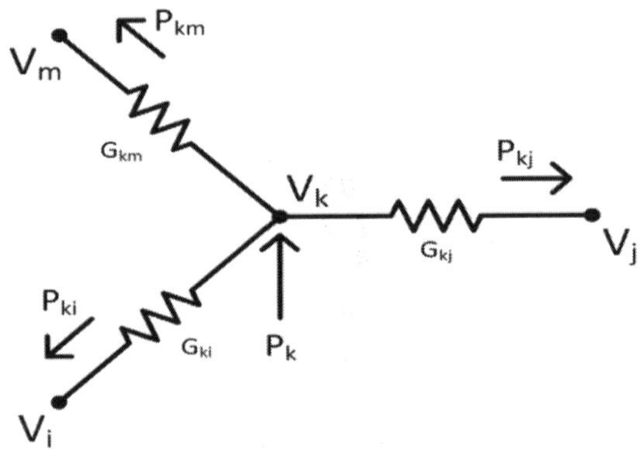

Figure 5. Power flow in a multi-bus system (based on Kirchhoff's law).

The Newton- Raphson method is implemented using Equation (4) to solve the DC power flow equation. The DC power flow Jacobian matrix is only composed of conductance components, and the Newton-Raphson iteration for DC power flow is performed as a function of the correlation between the voltage and the real power injection.

The objective of this paper is to estimate the optimal ESS charging schedule in order to maintain the voltage stability within the regulation range. Therefore, it is necessary to induce the formulation regarding the correlation between the voltage and the power injection at each PCC. From the DC power flow Equations (3) and (4), the voltage and power injection correlation equation are derived in a matrix form as in

Equations (5)–(7). This correlation matrix represents control of the power injection to adjust the distribution line voltages:

$$\Delta P_i = \sum_{k=1}^{N} \frac{\partial P_i}{\partial V_k} \Delta V_k \tag{5}$$

$$\begin{bmatrix} \Delta P_1 \\ \Delta P_2 \\ \Delta P_3 \\ \vdots \end{bmatrix} = \begin{bmatrix} \frac{\partial P_1}{\left(\frac{\partial V_1}{V_1}\right)} & \frac{\partial P_1}{\left(\frac{\partial V_2}{V_2}\right)} & \frac{\partial P_1}{\left(\frac{\partial V_3}{V_3}\right)} & \cdots \\ \frac{\partial P_2}{\left(\frac{\partial V_1}{V_1}\right)} & \frac{\partial P_2}{\left(\frac{\partial V_2}{V_2}\right)} & \frac{\partial P_2}{\left(\frac{\partial V_3}{V_3}\right)} & \cdots \\ \frac{\partial P_3}{\left(\frac{\partial V_1}{V_1}\right)} & \frac{\partial P_3}{\left(\frac{\partial V_2}{V_2}\right)} & \frac{\partial P_3}{\left(\frac{\partial V_3}{V_3}\right)} & \cdots \\ \vdots & \vdots & \vdots & \ddots \end{bmatrix} \begin{bmatrix} \frac{\Delta V_1}{V_1} \\ \frac{\Delta V_2}{V_2} \\ \frac{\Delta V_3}{V_3} \\ \vdots \end{bmatrix} \tag{6}$$

$$\frac{\partial P_i}{\left(\frac{\partial V_k}{V_k}\right)} = V_i V_k G_{ik} \tag{7}$$

3.2. Complementary Optimal ESS Scheduling for Voltage Regulation

3.2.1. Representation of the Complementary ESS Scheduling on the Basis of the Smart Metering Data

This paper presents an algorithm for the optimal operation of the ESS to regulate the distribution line voltage. The main contribution of this algorithm is the complementary operation between the ESSs. In order to complementally schedule all the ESSs in a distribution system, the power injection data at each PCC are required, and these data can be obtained by a smart metering system. The measured power injection data obtained by the smart meters at each PCC are transferred to a concentrator, which collects the entire distribution system information and estimates the current condition of the distribution system. From the power injection data, the distribution voltage can be derived as explained in the DC power flow calculation method. Therefore, the concentrator estimates the distribution voltage condition and provides an instruction on the control of the ESSs for the complementary operation between the ESSs [8]. Figure 6 represents the configuration of this distribution system, which is equipped with a smart meter system at each bus and a data concentrator.

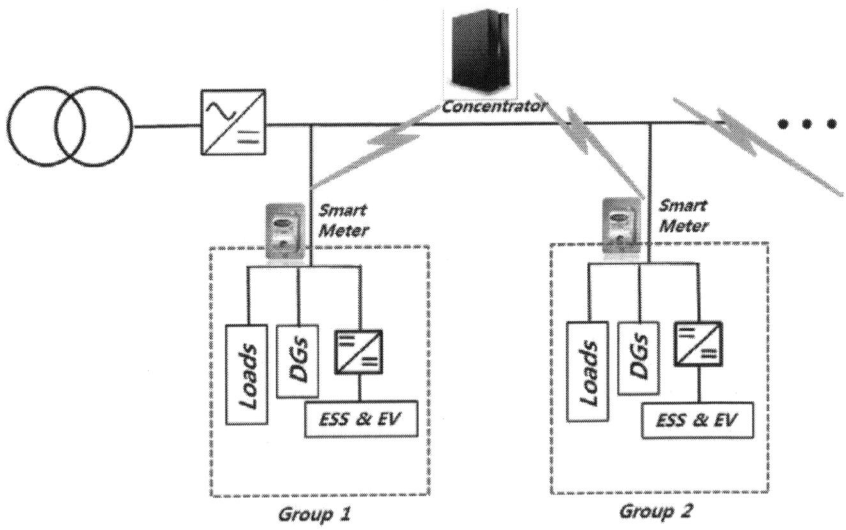

Figure 6. Configuration of the distribution system equipped with a smart metering system.

Case studies have been conducted with the real KEPCO distribution system data to demonstrate the ESS control algorithm. The Cheon-an distribution network in the KEPCO system comprises a roof-top PV generation at each house. In addition, six houses are lumped into one connection point, which is defined as the point of common coupling (PCC). This paper applies the OW-60 mm^2 single-phase pi (π) line parameter, which has a resistance, an inductance and a capacitance per unit length of 0.313 (Ω/km), 0.266 (mH/km), and 0.0013 (μF/km), respectively.

Each smart meter measures the loading condition at each PCC, and the voltage condition of the whole distribution system is estimated. By calculating the DC power flow on the basis of the measured information from the smart meter system, the distribution line voltage condition can be determined, whether the current voltage condition deviates from the normal range or not. When the estimated voltage condition is out of range, ESS charging is performed as instructed by the concentrator. The capacity of charging for each ESS is calculated by considering the mutual influence between the PCCs [9].

3.2.2. Complementally Optimized Scheduling for ESSs in a Distribution System

The distribution line voltage can be calculated by using the DC power flow calculation as explained in Section 3.1. In the case that the distribution system is radial, the power flow at bus i can be simplified as shown in Equation (8). The distribution line voltage is closely related to the power injection, and the individual power injection at each PCC affects not only the voltage of the connected PCC, but also the entire distribution line voltage. In this paper, in order to estimate the optimal ESS capacity for voltage regulation, the optimization function based on the genetic algorithm (GA) is used through MATLAB simulation [10,11]. In addition, the result of the algorithm is demonstrated through PSCAD/EMTDC simulation:

$$P_i = V_i\left(V_{i-1}G_{(i-1)i} + V_iG_{ii} + V_{i+1}G_{i(i+1)}\right) \tag{8}$$

During the periods of high DG generation, the voltage at a given PCC fluctuates above the permitted voltage regulation range. The amount of overvoltage can be calculated by performing the DC power flow calculation with the power injection data from obtained the smart meters, and the ESS can absorb an adequate power in order to maintain the voltage within the limits. As expressed in Equation (8), the voltage at a given PCC depends not only on the power usage at the connected PCC, but also on the condition of the adjacent bus with mutual influence. Therefore, when the ESS operation scheduling is estimated, the mutual influence between the PCCs must be considered.

By using the data from the smart metering system, this paper introduces a complementally optimized ESS scheduling algorithm for voltage regulation for all the ESSs, and not only for the individual ESS control independently. The required power, which can be generated or absorbed by the ESSs, is calculated for the voltage regulation (388 V $< V_{dc} <$ 412 V).

The presented ESS scheduling algorithm considers all the available ESSs in the distribution system to regulate the distribution line voltage; therefore, the objective function of optimization is to minimize the capacity of all the ESSs with satisfaction under the normal voltage operating condition as expressed in Equations (9)–(11):

$$basic\ function:\ P^{PCC} = (P_{DG} - P_L)^{scheduled} + P_{ESS} \quad (9)$$

$$objective\ function:\ \sum_{k=2}^{N} \left(P_k(V_{k-1}, V_k, V_{k+1}) - P_k^{scheduled}\right)^2 \quad (10)$$

with voltage regulation constraints:

$$388\ \text{V} < V_k < 412\ \text{V}\ (k = 1{\sim}N) \quad (11)$$

By performing an iterative GA process, the best result is stochastically selected. The optimized P for each ESS is calculated while monitoring whether the distribution line voltages at every PCC satisfy the voltage regulation requirements. Through this process, the P^{new} value required for voltage regulation is updated as expressed in Equation (12). In addition, from this result, the capacity of each ESS is determined. Figure 7 shows the flow chart of this entire process:

$$P_i^{new} = P_i^{old}\left(1 + \frac{\Delta P_i^{old}}{P_i^{old}}\right) \quad (12)$$

The algorithm is divided into three sections. The first section is the part where the system configuration is entered, and the power injection data of each PCC are collected from the smart metering system. In the second section the DC power flow calculation is performed on the basis of the collected power injection data. Through the DC power flow calculation, it can be estimated whether the distribution line voltage at each PCC is within the normal operation range or not. The third section involves the ESS scheduling process. In this part, the optimization using the GA is performed to determine the optimal operation condition for each ESS.

3.2.3. Methodologies for the Discharging Schedule of ESSs in a Distribution System

In order to regulate the voltage during the periods of high PV generation by using the ESSs, the state of charging (SOC) of the ESSs should be considered to ensure an adequate capacity of the ESSs. In order to ensure the capacity of an ESS for voltage regulation, the SOC of the ESSs must be lowered by adequately discharging the power during the relatively low DG generation and high load periods. The discharge power from the ESSs can cause the distribution line voltage fluctuation. This paper suggests two different methods in order to adequately discharge the ESS power. The main objective of this discharging algorithm is to enable the distribution line voltage operate closer to the normal voltage.

The first method is based on the historical power flow tendency. In this case, the charged power during the periods of high PV generation is discharged during historically low DG generation and high load periods. In this paper, it is assumed that the discharging periods are between 7 and 11 pm, and the charged energy is equally discharged during those periods. When the ESS SOC reaches the lowest level, the discharge operation is stopped.

The second method optimally discharges the available ESS power to increase the distribution line voltage for it to be closer to the normal voltage. This is performed through simulation by increasing the lower voltage limit during the optimization process. In this simulation, the lower voltage limit for the optimization is set as 393 V. Therefore, when the distribution line voltage reaches below 393 V, ESS discharging is complementally performed to compensate the voltage with optimal use of the ESS power.

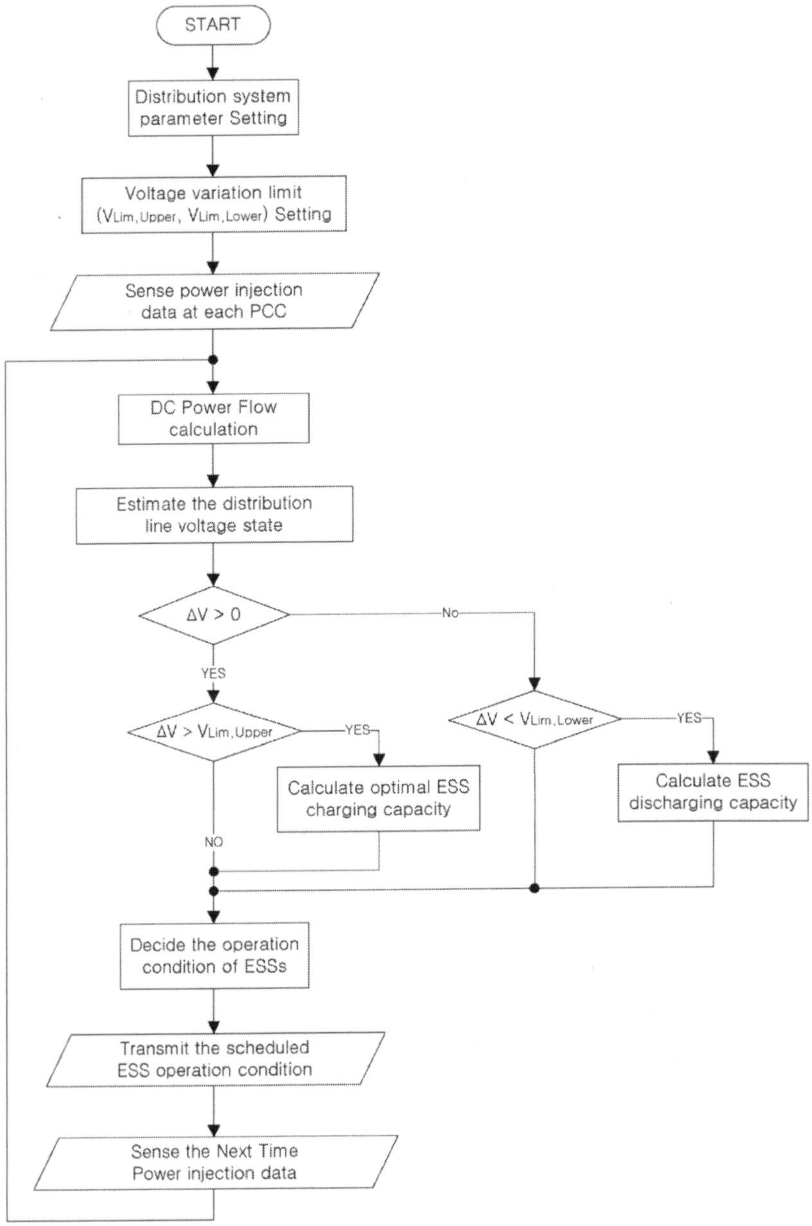

Figure 7. ESS charge scheduling algorithm for voltage regulation.

4. SIMULATION RESULTS

In order to estimate the optimal ESS capacity at each PCC for voltage regulation, MATLAB simulation is performed with the real KEPCO distribution data. The MATLAB simulation result is implemented through a PSCAD/EMTDC simulation for demonstration. In this paper, it is assumed that the PV penetration can be increased from the current 2.5-kW capacity in order to verify the ESS effect under the overvoltage condition. Furthermore, in order to verify the influence of the PV distribution state in a distribution system, four case studies are conducted: equal distribution and clumped distribution on the upper side, middle side, and lower side of the distribution line. Figure 8 shows the profile of the input data for the load demands at six PCCs, and Figure 9 represents the PV output generation for one day.

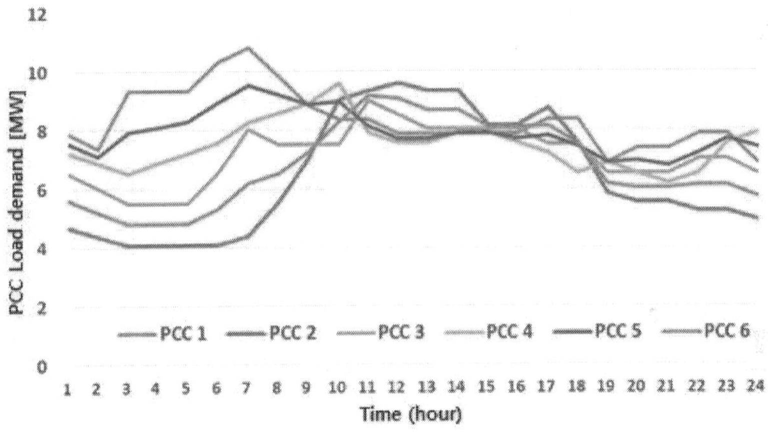

Figure 8. Load profiles for one day at six PCCs (each PCC consists of six houses).

4.1. Complementally Optimized ESS Operation Scheduling for Voltage Regulation

As shown in Figure 10, when the PV generation is greater than 4.0 kW, some of the PCCs on the distribution line experience overvoltage problems. In order to verify the performance of the ESS scheduling algorithm for voltage regulation, this algorithm is applied to the Cheon-an distribution system that is composed of 36 houses, each having an increased PV capacity 4.0-kW peak. The DC power flow calculation is performed to estimate the ESS capacity for voltage regulation, and the calculation result is summarized in Table 2.

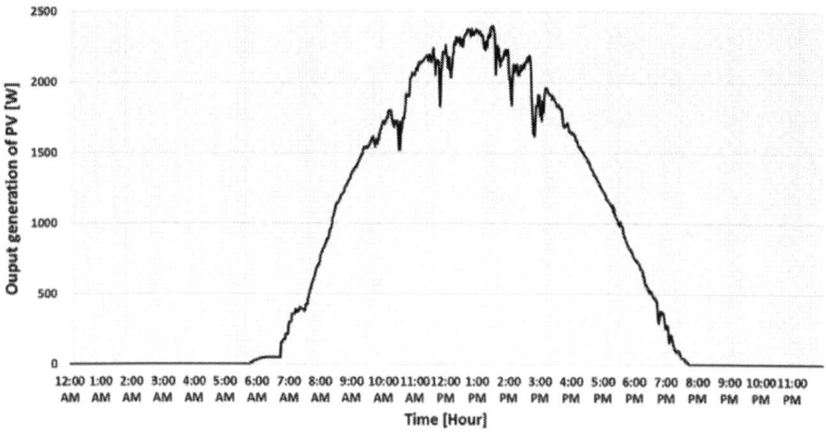

Figure 9. Output generation of a PV on a house (2.5-kW PV).

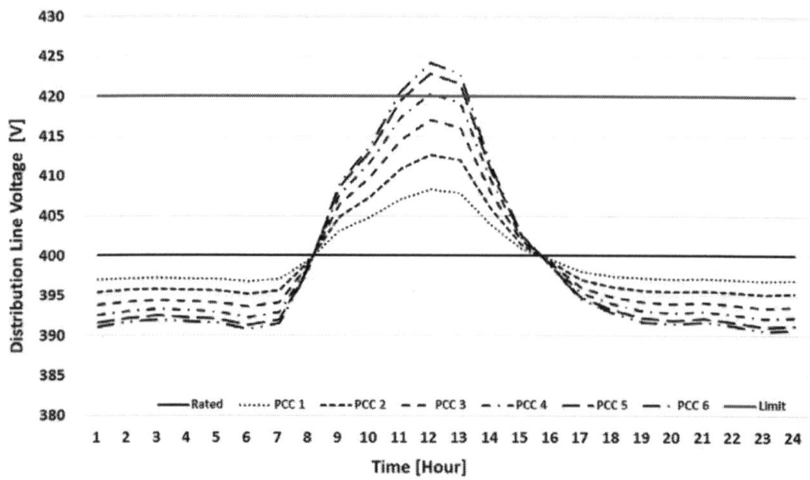

Figure 10. Voltage profile for one day (PV = 4 kW).

Table 2. ESS charging schedule and capacity for voltage regulation (4.0-kW PV is equally distributed).

Time	PCC 1	PCC 2	PCC 3	PCC 4	PCC 5	PCC 6	Total
11:00–12:00	124 W	202 W	296 W	390 W	501 W	614 W	31 kW·h
12:00–13:00	987 W	1609 W	2362 W	3115 W	4011 W	4920 W	
13:00–14:00	666 W	1086 W	1594 W	2102 W	2706 W	3318 W	
Capacity	1777 W·h	2897 W·h	4254 W·h	5608 W·h	7219 W·h	8853 W·h	

As shown in Figure 11, by using this ESS control algorithm, the distribution line voltage can satisfy the voltage regulation requirement. The main objective of this paper is to propose an ESS control algorithm by considering the complementary operation among all the ESSs, and not the individual ESS control. With the presented algorithm, the optimal ESS charging schedule which satisfies the voltage regulation requirement is determined. As depicted in Figure 12, the ESSs at PCCs 1–3, which do not experience overvoltage problems, are also charged to regulate the voltage of the other PCCs. With the presented algorithm, it is verified that the ESSs are charged to operate complementarily even when the adjacent PCC does not experience the voltage problem.

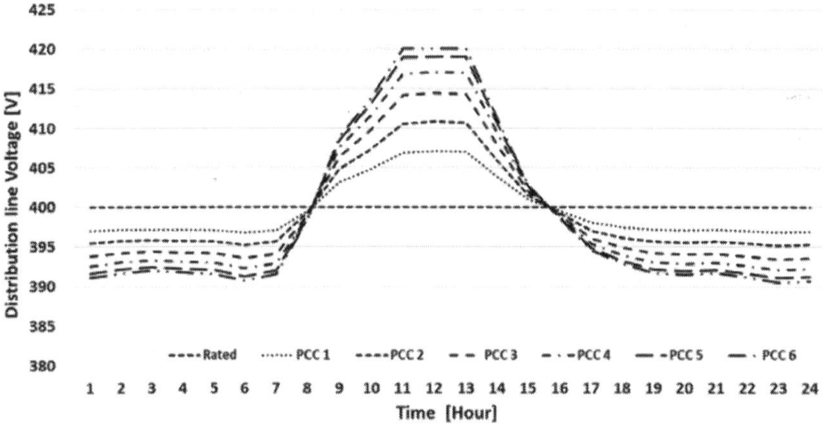

Figure 11. Voltage profile compensated by the ESSs.

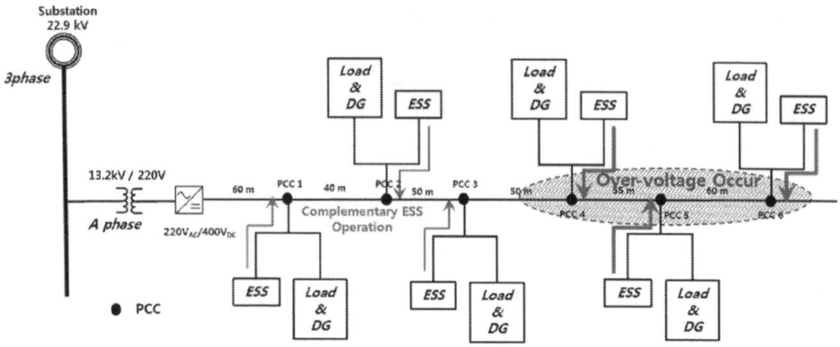

Figure 12. Complementary ESS operation for voltage regulation.

4.2. Comparison of the ESS Scheduling with Different PV Distributions in the Distribution Line

For the analysis of the influence of the PV distribution state on the ESS capacity, three case studies are conducted in different PV distribution states. Table 3 summarizes the charging schedule for the case that most of the PVs are connected on the upper side of the distribution line, PCCs 1–3. Table 4 summarizes the case that most of the PVs are connected on the lower side of the distribution line, PCCs 4–6, and Table 5 summarizes the case that most of the PVs are connected on the middle side. The total sum of the PV capacities is the same for all the cases.

From this result, it is verified that all the ESSs in the distribution line operate complementally to minimize the total ESS capacity. As can be observed from the simulation result, a high ESS capacity is required on the lower side of the distribution line. Furthermore, a high-capacity PV connected on the lower side of the distribution line requires a high ESS capacity. Therefore, when the ESS installation is considered in a distribution line, it is necessary to consider efficient distribution of the ESSs in advance.

Table 3. ESSs charging schedule and capacity (case 2: most of the PVs are distributed on the upper side).

Time	PCC 1	PCC 2	PCC 3	PCC 4	PCC 5	PCC 6	Total
11:00–12:00	637 W	1,038 W	1,525 W	2,013 W	2,594 W	3,182 W	59 kW·h
12:00–13:00	1,550 W	2,528 W	3,716 W	4,909 W	6,330 W	7,768 W	
13:00–14:00	1,204 W	1,962 W	2,884 W	3,810 W	4,912 W	6,027 W	
Capacity	3,392 W·h	5,529 W·h	8,126 W·h	10,734 W·h	13,837 W·h	16,978 W·h	

Table 4. ESSs charging schedule and capacity (case 3: most of the PVs are distributed on the lower side).

Time	PCC 1	PCC 2	PCC 3	PCC 4	PCC 5	PCC 6	Total
11:00–12:00	1,377 W	2,245 W	3,295 W	4,340 W	5,586 W	6852 W	99 kW·h
12:00–13:00	2,375 W	3,873 W	5,686 W	7,491 W	9,646 W	11,834 W	
13:00–14:00	1,987 W	3,240 W	4,756 W	6,265 W	8,066 W	9,895 W	
Capacity	5,740 W·h	9,358 W·h	13,738 W·h	18,098 W·h	23,299 W·h	28,581 W·h	

Table 5. ESSs charging schedule and capacity (case 4: most of the PVs are distributed on the middle side).

Time	PCC 1	PCC 2	PCC 3	PCC 4	PCC 5	PCC 6	Total
11:00–12:00	976 W	1,590 W	2,332 W	3,076 W	3,969 W	4,872 W	77 kW·h
12:00–13:00	1,928 W	3,140 W	4,607 W	6,081 W	7,853 W	9,645 W	
13:00–14:00	1,562 W	2,544 W	3,733 W	4,927 W	6,361 W	7,812 W	
Capacity	4,466 W·h	7,275 W·h	10,673 W·h	14,085 W·h	18,185 W·h	22,330 W·h	

4.3. ESS Discharging Schedule Simulation

As shown in Figure 13, discharging the ESS power can change the voltage profile. By discharging the power at an appropriate time when the load is high and DG generation is low, the voltage can be compensated to be closer to the normal voltage. In this paper, two different algorithms are proposed to appropriately discharge the ESS power. The proposed discharging algorithms enable the distribution line voltage operate closer to the normal voltage. The simulated case is based on the system data which is defined as equally distributed case in Section 4.1.

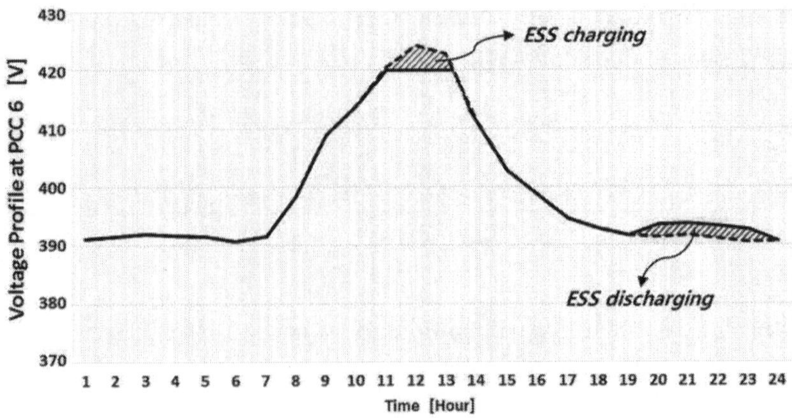

Figure 13. Voltage compensation by using the ESSs.

Table 6 summarizes the result when the available ESS power is equally discharged from 7 to 11 pm. In this case, it is assumed that this time period corresponds to the time when the load demand is low and DG generation is high. This discharging period must be set before the discharging is performed. In addition, this time period can be varied according to the distribution system load and generation configuration.

Table 6. ESS equally discharging schedule between 7 and 11 pm.

Time	PCC 1	PCC 2	PCC 3	PCC 4	PCC 5	PCC 6
19:00–20:00	444 W	724 W	1063 W	1402 W	1804 W	2213 W
20:00–21:00	444 W	724 W	1063 W	1402 W	1804 W	2213 W
21:00–22:00	444 W	724 W	1063 W	1402 W	1804 W	2213 W
22:00–23:00	444 W	724 W	1063 W	1402 W	1804 W	2213 W

The other algorithm for discharging the power is based on the optimization technique. Through the DC power flow calculation, the distribution line voltage can be estimated. The capacity of discharging power of each ESS can be calculated with the GA method. By setting the voltage lower limit, when the distribution line voltage reaches below the defined lower voltage limit, the complementary optimized discharging capacity of each ESS for voltage regulation is calculated. Table 7summarizes the result of this optimal discharging algorithm. In addition, this result shows that when the ESS is fully discharged, the discharging operation is complete.

Table 7. ESS optimally discharging schedule for voltage compensation.

Time	PCC 1	PCC 2	PCC 3	PCC 4	PCC 5	PCC 6
19:00–20:00	321 W	530 W	786 W	1042 W	1345 W	1646 W
20:00–21:00	276 W	456 W	676 W	897 W	1158 W	1417 W
21:00–22:00	374 W	616 W	914 W	3667 W	4715 W	5788 W
22:00–23:00	804 W	1293 W	1876 W	-	-	-

5. CONCLUSIONS

This paper introduced a voltage regulation concept for low-voltage grids with high PV penetration that use the ESS as the solution. Using a smart metering system, the algorithm presented in this paper performs the complementary operation of all the ESSs installed in a distribution line, and it is not a local control method for individual ESS systems. The power usage/generation data at each PCC is collected from the smart metering system and from these data, the complementary operation scheduling for all the ESSs is performed. Through a PSCAD simulation, it is verified that the derived result of the ESS scheduling is effective for voltage regulation. The required ESS capacity varies according to the system structure. It is expected that the proposed methodology and formulations will be useful for grid planning.

ACKNOWLEDGMENTS

This research was made possible by the support of the KEPCO (Korea Electric Power Corporation) project “Modeling and Analysis of LVDC Distribution Network and Equipment” (No. R12DA01) and by the Human Resources Development Program (No. 20134030200340) of the Korea Institute of Energy Technology Evaluation and Planning (KETEP) grant.

CONFLICTS OF INTEREST

The authors declare no conflict of interest.

REFERENCES

1 Pourmousavi, S.A.; Cifala, A.S.; Nehrir, M.H. Impact of High Penetration of PV Generation on Frequency and Voltage in a Distributed Feeder. In Proceedings of the North American Power Symposium (NAPS), Champaign, IL, USA, 9–11 September 2012.

2 Kim, T.E.; Kim, J.E. Voltage regulation coordination of distributed generation system in distribution system. *Power Eng. Soc. Summer Meet.* **2001**, *1*, 480–484.

3 Choi, J.H.; Kim, J.C. Advanced voltage regulation method of power distribution systems interconnected with dispersed storage and generation systems. *IEEE Trans. Power Deliv.* **2001**, *16*, 329–334.

4 Marra, F.; Yang, G.Y.; Traeholt, C.; Larsen, E.; Ostergaard, J.; Blazic, B.; Deprez, W. EV charging facilities and their application in LV feeders with photovoltaics. *IEEE Trans. Smart Grid* **2013**, *4*, 1533–1540.

5 Yu, W.; Lie, D.; Huang, Y. Operation optimization based on the power supply and storage capacity of an active distribution network. *Energies* **2013**, *6*, 6423–6438.

6 De Oliveira Gomes, A.; Meirelles Gouvea, M. DC Power Flow Optimization with a Parallel Evolutionary Algorithm. In Proceedings of the Transmission and Distribution: Latin America Conference and Exposition (T&D-LA), Montevideo, Uruguay, 3–5 December 2012.

7 Lu, S.; Zhou, N.; Kumar, N.P.; Samaan, N.; Chakrabarti, B.B. Improved DC Power Flow Method Based on Empirical Knowledge of the System. In Proceedings of the Transmission and

Distribution Conference and Exposition, New Orleans, LA, USA, 19–22 April 2010.

8 Chakraborty, S.; Seniyu, T.; Toyama, H.; Saber, A.Y.; Funabashi, T. Determination methodology for optimizing the energy storage size for power system. *Gener. Trans. Distrib. IET* **2009**, *3*, 987–999.

9 Depuru, S.S.S.R.; Wang, L.; Devabhaktuni, V. Smart meters for power grid: Challenges, issues, advantages and status.*Renew. Sustain. Energy Rev.* **2011**, *15*, 2736–2742.

10 Birge, B. PSOt-a Particle Swarm Optimization Toolbox for Use with MATLAB. In Proceedings of the 2003 IEEE Swarm Intelligence Symposium, Indianapolis, IN, USA, 24–26 April 2003.

11 Glavin, M.E.; Chan, P.K.W.; Hurley, W.G. Optimization of Autonomous Hybrid Energy Storage System for Photovoltaic Applications. In Proceedings of the Energy Conversion Congress and Exposition, San Jose, CA, USA, 20–24 September 2009.

CHAPTER 7

Operation Optimization Based on the Power Supply and Storage Capacity of an Active Distribution Network

Wenpeng Yu, Dong Liu * and Yuhui Huang

Key Laboratory of Control of Power Transmission and Conversion, Ministry of Education, Department of Electrical Engineering, Shanghai Jiao Tong University, Shanghai 200240, China

ABSTRACT

Due to the interconnection and active management of Distributed Generation (DG) and Energy Storage Systems (ESSs), the traditional electrical distribution network has become an Active Distribution Network (ADN), posing challenges to the operation optimization of the network. The power supply and storage capacity indexes of a Local Autonomy Control Region (LACR), which consists of DGs, ESSs and the network, are proposed in this paper to quantify the power regulating range of a LACR. DG/ESS and the network are considered as a whole in the model of the indexes, considering both network constraints and power constraints of the DG/ESS. The index quantifies the maximum LACR power supplied to or received from ADN lines.

Similarly, power supply and storage capacity indexes of the ADN line are also proposed to quantify the maximum power exchanged between ADN lines. Then a practical algorithm to calculate the indexes is presented, and an operation optimization model is proposed based on the indexes to maximum the economic benefit of DG/ESS. In the optimization model, the power supply reliability of the ADN line is also considered. Finally, the indexes of power supply and storage capacity and the optimization are demonstrated in a case study.

KEYWORDS

distributed generation; energy storage system; renewable energy; active distribution network; local autonomy control region

1. INTRODUCTION

Various kinds of renewable energy have been so far integrated into electrical power distribution networks in the form of Distributed Generation (DG) systems to reduce societys' dependence on fuel resources. Meanwhile, Energy Storage Systems (ESSs) are also connected to the network. The grid-connected DGs and ESSs greatly change the characteristics of the network and complicate the operation optimization, and the intermittence of renewable energy poses risks to the operation of the whole network. To address these issues, active management of the network and DGs/ESSs has been broadly utilized, suppressing the intermittence of the renewable energy and reducing the peak-valley differences of the loads. The active management also puts off the upgrade of power facilities [1,2], improves the power supply reliability, and results in the conversion of the traditional distribution network into an Active Distribution Network (ADN).

To date, abundant research have been done on the active management and operation optimization of grid-connected DGs/ESSs. Some forms of renewable energy, e.g., wind generation and photovoltaic arrays, *etc.*, are intermittent and thus increase the fluctuation of the net load of the ADN lines. High power density ESS [3,4] has been utilized to suppress the short-term power fluctuation, while high energy density ESS is utilized to smooth the renewable energy power in the long term [5,6], as well as reduce the peak-valley load differences of ADN lines [7,8]. Hybrid storage systems, consisting of both high power density ESS and high energy density ESS, by combining the merits from these two types of the ESS, are regarded as more advantageous [9,10,11]. The real-time control of ESSs suppresses the impact of intermittent energy sources on the network, however, operation optimization of the whole network should be considered in the real-time control of ESS. Different ADN optimization models considering the line losses or electricity costs are presented in [12,13], and optimization strategies are discussed in [14,15] for the operation of combined heating, cooling and power systems. Since the powers of individual DGs/ESSs are the variables that need to be optimized in the optimization, the computation cost increases rapidly with the number of DGs/ESSs considered. Moreover, various kinds of DGs, ESSs, loads and networks are combined into a system, which is able to actively regulate its power. The power regulating ability is one of the key distinctions of ADNs from traditional distribution networks, and this has not been well studied yet.

In this work, the power supply and storage capacity of the network is analyzed, and a long-term operation optimization model is proposed based on the capacity to maximize the DG/ESS economic benefits. Within an ADN line, the Local Autonomy Control Region (LACR), which consists of DGs/ESSs, loads and network, is operated as a whole. The power supply & storage capacity index of the LACR is proposed to quantify the maximum power regulation range of the LACR, considering both the DG/ESS power constraints and the network restrictions. The power supply and storage capacity index of an ADN line is also presented to quantify the maximum power exchanged between adjacent ADN lines through tie switches. Practical algorithms are used to calculate the power supply and storage indexes, and an operation optimization model of an ADN line based on the indexes is proposed to maximize the economic benefits of DG/ESS. The power supply reliability of the ADN line is also considered in the optimization. Finally the indexes of power supply and storage capacity and the optimization are demonstrated in a case study.

2. POWER SUPPLY & STORAGE CAPACITY OF LOCAL AUTONOMY CONTROL REGION

According to the survey of the CIGRE C6.11 working group, active distribution networks are distribution networks that have systems in place to control a combination of Distributed Energy Resources (DERs) [16], and have the ability to manage the electricity flows. With the active management of DERs and electricity flows, feeders of the traditional distribution network are converted to ADN lines. Within an ADN line, part of the DERs and loads make up a Local Autonomy Control Region which is managed and operated as a whole to reduce the optimization computation cost and relieve the communication pressure between the control center and DERs [17,18]. As shown in Figure 1, three LACRs exist in ADN line A and line B. The area of the LACR is generally specified according to the topological connection and the allocation of the autonomic controller. The branch with an autonomic controller and more than one DER is regard as a LACR. The autonomic controller receives the control target of the whole LACR from the control center, and coordinates the operation of DERs within the LACR, according to that control target.

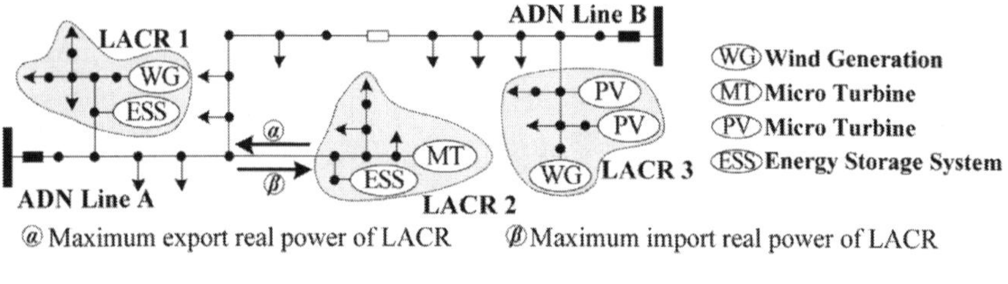

Figure 1. ADN line and local autonomy control region.

The maximum export/import power of LACR needs to be calculated so that the LACR could be regarded as a whole in the operation optimization of the ADN line. The power supply and storage capacity of the LACR is defined as the maximum export/import real power of the LACR, considering power constraints of DGs/ESSs as well as any restrictions of node voltage and maximum branch current. The power supply capacity of the LACR is the maximum export power (power α inFigure 1), whose positive direction is defined as injecting into an ADN line, while the power storage capacity is the maximum import power of LACR (power β in Figure 1), whose positive direction is defined as injecting into the LACR. The power constraints of DGs/ESSs, node voltage restrictions, branch current and SOC of ESSs are considered in the calculation of the capacity, as discussed below.

2.1. Power Constraints of Distributed Generation

DGs can be classified into two categories according to their power controllablity: power controllable DGs and uncontrollable DGs. The power of biomass power generators, micro gas turbines or fuel cells is controllable, while the wind generation and photovoltaic cell power is intermittent and dependent on the weather. The power restrictions of the two types of DG are shown in Equation (1):

$$\begin{cases} P_i^{\min} \le P_i \le P_i^{\mathrm{R}}, & i \in \mathbf{F}_{\mathrm{ctrl}} \quad \text{(a)} \\ P_j = P_j^{\mathrm{E}}, & j \in \mathbf{F}_{\mathrm{inte}} \quad \text{(b)} \end{cases} \tag{1}$$

where $\mathbf{F}_{\mathrm{ctrl}}$ is the set of power controllable DGs; P_i is the power of the ith power-controllable DG; while $P_i^{\min}$ and P_i^{R} are the minimum operating power and the rated power; $\mathbf{F}_{\mathrm{inte}}$ is the set of power uncontrollable DGs; P_j is the power of j^{th}power-uncontrollable DG; and P_j^{E} is the expected maximum power according to the weather forecast. Some of the intermittent renewable energy generators have the ability to adjusting the power by changing the operating point, but in order to maximize the exploitation of renewable energy, the renewable energy generators are assumed to be operated at their maximum power point according to the power forecasting.

2.2. Power Constraints of Energy Storage System

The charging/discharging power of an ESS is subjected to both constraints of rated power and constraints of the ESS' SOC. The change of SOC during charging/discharging is calculated by Equation (2):

$$\Delta S_{oc}(\Delta T) = \begin{cases} \int_{T}^{T+\Delta T} [\mu P_c(t)/E_R] \cdot dt & \text{charging} \\ -\int_{T}^{T+\Delta T} \frac{P_d(t)}{\eta \cdot E_R} \cdot dt & \text{discharging} \end{cases} \tag{2}$$

where $\Delta S_{oc}(\Delta T)$ is the change of SOC during time period of $[T, T+\Delta T]$; μ and η are charging/discharging efficiencies respectively; P_c and P_d are charging/discharging power; E_R is the rated energy capacity of the ESS.

As excessive charging or discharging may cause damage to the ESS, Equation (3) should be satisfied:

$$S_{oc}^{\min} \le S_{oc}(T) \le S_{oc}^{\max} \tag{3}$$

where $S_{oc}(T)$ is the SOC at time T; while $S_{oc}^{\max}$ and $S_{oc}^{\min}$ are the upper and lower limitation of SOC respectively. After the charging/discharging in the time interval $[T, T+\Delta T]$, the value of SOC at $T+\Delta T$ should satisfy the Inequation (3). Then Inequation (4) is derived according to Equations (2) and (3), ignoring the power changes during $[T, T+\Delta T]$. The symbols in Equation (4) have the same meaning with those in Equations (2) and (3):

$$\begin{cases} P_c \leq \dfrac{S_{oc}^{\max} - S_{oc}(T)}{\mu \cdot \Delta T} \cdot E_R \\ P_d \leq \dfrac{S_{oc}(T) - S_{oc}^{\min}}{\Delta T} \cdot \eta \cdot E_R \end{cases} \tag{4}$$

The power constraints of ESS are shown as Equation (5), considering constraints of both SOC and rated charging/discharging power. P_c^R and P_d^R are the rated charging/discharging power, $P_c^{\max}$ and $P_d^{\max}$ are the maximum charging/discharging power calculated according to the constraints:

$$\begin{cases} P_c \leq P_c^{\max} = \min\left[\dfrac{S_{oc}^{\max} - S_{oc}(T)}{\mu \cdot \Delta T} \cdot E_R, P_c^R\right] \\ P_d \leq P_d^{\max} = \min\left[\dfrac{S_{oc}(T) - S_{oc}^{\min}}{\Delta T} \cdot \eta \cdot E_R, P_d^R\right] \end{cases} \tag{5}$$

2.3. The Model of Power Supply & Storage Capacity of Local Autonomy Control Region

The power supply and storage capacity of a LACR is the maximum power that the LACR can supply to or receive from an ADN line, considering both the power constraints of the DG/ESS and constraints of node voltages and branch currents. The model of the capacity is given by Equations (6)–(8), in which, Equations (6) and (7) are the objective functions and Equation (8) are the constraints:

$$P_{\mathrm{su}}^{\max} = \max\left(\sum_{i \in \mathbf{F}_{\mathrm{ctrl}}} P_i + \sum_{j \in \mathbf{F}_{\mathrm{inte}}} P_j^{\mathrm{E}} + \sum_{k \in \mathbf{F}_{\mathrm{ESS}}} P_{d.k} - \sum_{l \in \mathbf{F}_L} P_l\right) \tag{6}$$

$$P_{st}^{max} = \max(-\sum_{i \in \mathbf{F}_{ctrl}} P_i - \sum_{j \in \mathbf{F}_{inte}} P_j^E + \sum_{k \in \mathbf{F}_{ESS}} P_{c.k} + \sum_{l \in \mathbf{F}_L} P_l) \tag{7}$$

$$\begin{cases} V_{min} \leq V \leq V_{max} & \text{(a)} \\ I_b \leq I_{b.Rated} & \text{(b)} \\ P_i^{min} \leq P_i \leq P_i^R, i \in \mathbf{F}_{ctrl} & \text{(c)} \\ P_{c.k} \leq P_{c.k}^{max}, k \in \mathbf{F}_{ESS} & \text{(d)} \\ P_{d.k} \leq P_{d.k}^{max}, k \in \mathbf{F}_{ESS} & \text{(e)} \end{cases} \tag{8}$$

In Equations (6) and (7), P_{su}^{max}, P_{st}^{max} are the power supply and storage capacity to be calculated. The computations of "∑", "+", "−" are calculated by the power flow method, in which network losses are considered. $\mathbf{F}_{ESS}$ is the set of ESSs inside the LACR, and $\mathbf{F}_L$ is the set of loads. $P_{c,k}$ and $P_{d,k}$ are the charging and discharging power of the kth ESS. P_l is the power ofl^{th} load.

Equation (8a) is the node voltage restriction, where V_{max} and V_{min} are the upper and lower node voltage limitations. Equation (8b) is the branch current restriction, where $I_{b,Rated}$ is the rated maximum current of the bth branch. Equation (8c) is the power constraint of power controllable DGs, and Equations (8d,e) are charging/discharging power constrains of ESSs.

The positive direction of the power supply capacity of the LACR is defined as power injected into the ADN line, while that of the power storage capacity is defined as power received from the ADN line. A LACR may have the characteristics of power controllable DGs, power controllable loads or storage systems, according to the value of its power supply and storage capacity.

As shown in Figure 2, positive power supply capacity and negative storage capacity indicate that the maximum power supplied by DGs/ESSs in the LACR is greater than the power of loads, and the LACR has the characteristic of power controllable DGs. On the contrary, negative power supply capacity and positive storage capacity means that the maximum power supplied by DGs/ESSs is smaller than the power of loads, and the LACR has the power controllable load characteristics. Besides, if both power supply capacity and storage capacity are positive, the power regulating range of the LACR is large, and the LACR is regarded as having the energy storage system characteristics.

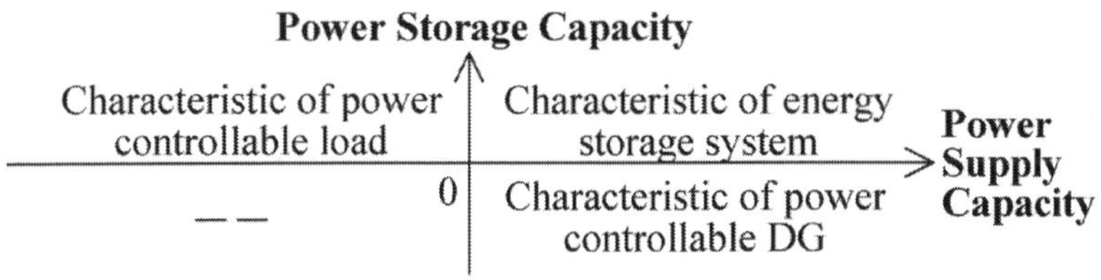

Figure 2. Characteristics of a LACR according to the power supply/storage capacity.

3. THE POWER SUPPLY & STORAGE CAPACITY OF ADN LINE

The power supply and storage capacity of an ADN line denotes the maximum power exchanged between adjacent ADN lines, as shown in Figure 3. In Figure 3a, since a fault occurs in line B, the load of line B is transferred to line A by closing tie switch I-II, and the power supply capacity of line A indicates the maximum power supplied to the ADN line B through the two tie switches. Supposing excessive renewable energy exists in line B, as shown in Figure 3b, then in order to maximize the exploitation of the renewable energy, the network is reconfigured and the excessive renewable energy of line B is consumed or stored by line A. The power storage capacity of line A indicates the maximum power injected into line A through tie switch I-II by line B.

In Figure 3 there is more than one tie switch between the ADN lines, and the power supply and storage capacity is the maximum summation of the power transmitted by all the tie switches. If there is no adjacent line, the power supply and storage capacity of ADN line is 0.

The model of power supply and storage capacity of ADN line is shown in Equations (9)–(11):

$$P_{\text{su.N}}^{\max} = \max(P_B + \sum P_m - \sum P_l) \tag{9}$$

$$P_{\text{st.N}}^{\max} = \max(-P_B - \sum P_m + \sum P_l) \tag{10}$$

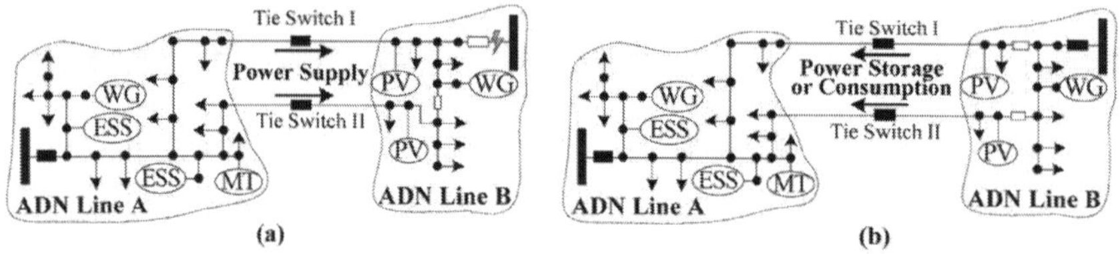

Figure 3. (**a**) Power supply capacity of ADN line; (**b**) Power storage capacity of ADN line.

Subject to:

$$\begin{cases} V_{\min} \le V \le V_{\max} & \text{(a)} \\ I_b \le I_{b.Rated} & \text{(b)} \\ -P_{\text{st.m}}^{\max} \le P_m \le P_{\text{su.m}}^{\max} & \text{(c)} \\ P_B \ge 0 & \text{(d)} \end{cases} \tag{11}$$

where $P_{\text{su.N}}^{\max}$ and $P_{\text{st.N}}^{\max}$ are the supply capability and storage capability of ADN line; P_B is the power injected into ADN line by high-voltage distribution network; P_m is the power of mth LACR, whose positive direction is defined as injecting into the ADN line; P_i is the power of the loads outside the LACR. Equations (11a,b) are the constraints of node voltage and branch current. Equation (11c) is the power restriction of the LACR, and $P_{\text{su.m}}^{\max}$, $P_{\text{st.m}}^{\max}$ are the power supply capability and storage capability of the mth LACR, respectively, which are calculated by Equations (6) and (7). Equation (11d) is the power constraint of the high-voltage grid. The power of the high-voltage grid injected into ADN line should be positive. The network loss is considered in the computation of "∑", "+" and "–". Other symbols in Equations (9)–(11) have the same meaning as those in Equations (6)–(8).

4. ALGORITHM OF THE POWER SUPPLY & STORAGE CAPACITY OF THE LACR AND ADN LINE

The power supply and storage capacity of the LACR and ADN line are proposed in order to reduce the calculation costs and complexity of the operation optimization of DGs/ESSs and ADN lines. Therefore the algorithm used for calculating the capacity should be practical.

Both the node voltage and branch current of the network will change with the DG/ESS power. As the constraints of node voltage and branch current are considered in the model of power supply and storage capacity, the change of voltage and current caused by DGs/ESSs need to be quantified.

The voltage deviation rate [19] of the network is defined in Equation (12), which indicates the voltage deviation of all the nodes in the network from the nominal voltage:

$$RD = 1 - \frac{\sum_{i=1}^{M}\left[(U_i - U_{\min})(U_{\max} - U_i)|P_i|\right]}{(U_{nom} - U_{\min})(U_{\max} - U_{nom}) \cdot \sum_{i=1}^{M}|P_i|} \tag{12}$$

where Ui and P_i are the voltage and real power of ith node; M is the count of nodes in the network; U_{nom} is the nominal voltage, while U_{min} and U_{max} are the lower and upper limitation of the voltage according to the voltage quality standard. U_{nom}, U_{min} and U_{max} are constants. Taking a medium distribution network for instance, Unom=10 kV, Umin=10 kV×95% and Umax=10 kV×105%. $|P_i|$ is regarded as the weight of the voltage deviation of the ith node. The smaller the voltage deviation of the nodes in the network, the smaller the value of *RD*, and *RD* will be 0 if the voltage of all the nodes is equal to the nominal voltage. The loading rate of the network is defined in Equation (13), which indicates the mean loading rate of all the cables and overhead conductors in the network:

$$RL = \frac{1}{B}\sum_{j=1}^{B}\frac{I_j}{I_{j.\max}} \tag{13}$$

where I_j is the current of jth conductor; and $I_{j,max}$ is the rated maximum current of the conductor; *RL* is the mean loading rate of the whole network.

The value of *RD* and *RL* will change with the power of DG/ESS, and the changes of *RD* and *RL* are regards as the impact of DG/ESS on the network. The impact factor of kth DG/ESS is defined as Equation (14), in which, ΔP_k is the power change of the kth DG/ESS, ΔRD and ΔRL denote the changes of *RD* and *RL* caused by ΔP_k:

$$C_k = (\Delta RD + \Delta RL)/\Delta P_k \tag{14}$$

The impact factor is different for different DGs/ESSs. The value of C_k is dependent on the topological connection of the DGs/ESSs and the branch current of the network.

A stepwise optimization method is implemented to calculate the power supply and storage capacity. The power of DG/ESS is initialized to 0, then a loop is started, in which the impact factors os DGs/ESSs on the network are calculated and the power of the DG/ESS with the minimum impact factor is increased by one power step. Constraints are verified in the loop, and the loop will stop if the constraints are not satisfied. The algorithm processes for power supply and storage capacity of the LACR and ADN line are discussed below.

4.1. Algorithm of Power Supply and Storage Capacity of the LACR

In the proposed algorithm procedure, the DGs/ESSs in LACRs are sorted according to their impact factor on the network, and the power of the DG/ESS with the minimum impact factor is increased by one power step. The constraints in Equation (8) are also verified in the loop. The algorithm procedure of the power supply capacity of LACR is described as follows:

(1) Initialization. The power of all DGs/ESSs in the LACR is initialized to 0, and the upper and lower limitation of DGs'/ESSs' power are calculated according to the power constraints discussed in Section 2.1 and Section 2.2. A list is created which consists of all the DGs/ESSs inside the LACR, and the loop starts.

(2) The impact factors of each DG/ESS are calculated according to Equation (14).

(3) The DGs/ESSs in the list created in step (1) are sorted according to their impact factors. The power of the DG/ESS with the minimum impact factor is increased by one power step.

(4) Constraints Verification. The constraints stated in Equation (8) are verified in the loop. If any constraint cannot be satisfied, cancel the power increment in step (3), and remove the DG/ESS from the list.

(5) If all the DGs/ESSs are removed from the list, the loop stops and go to step (6), else go to step (2).

(6) Power flow algorithm is executed to calculate the total power of the LACR supplied to the ADN line which is the power supply capacity of the LACR.

The block diagram of the power supply capacity calculation of the LACR is depicted in Figure 4. The calculation process of the power storage capacity of the LACR is similar to that of the power supply capacity, while the different exists in the step (3): the power of DG/ESS with the minimum impact factor decreases by one step or the discharging power increases.

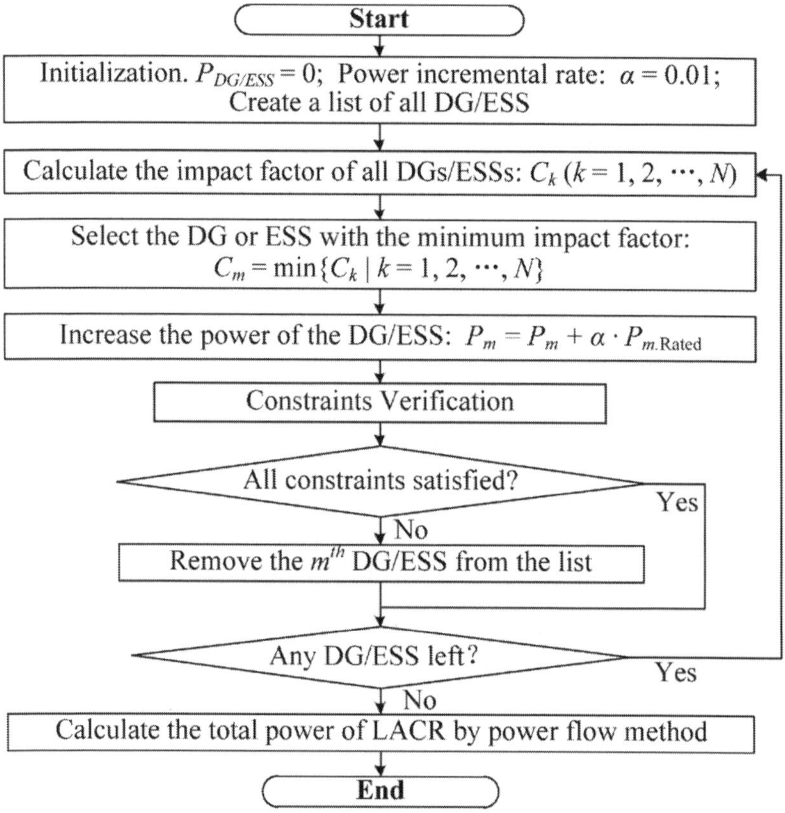

Figure 4. Block diagram of calculating the power supply capacity of the LACR.

4.2. Algorithm of the Power Supply & Storage Capacity of ADN Line

The power supply capacity of the ADN line is equal to the maximum power supplied to an adjacent line through a closed tie switch. In order to calculate the power supply capacity of an ADN line, the adjacent line is assumed to be a virtual power-controllable load, and P_V is used to represent the power of the virtual load. During the iterations, the value of P_V increases step by step and the constraints stated in Equation (11) are verified. If any of the constraints is not satisfied, the loop stops and the value of power supply capacity of ADN line is equal to P_V. If there is more than one tie switch between the two ADN lines, the adjacent line is assumed to be the same amount of virtual power-controllable load, and the power supply capacity is equal to the maximum total power of all virtual loads.

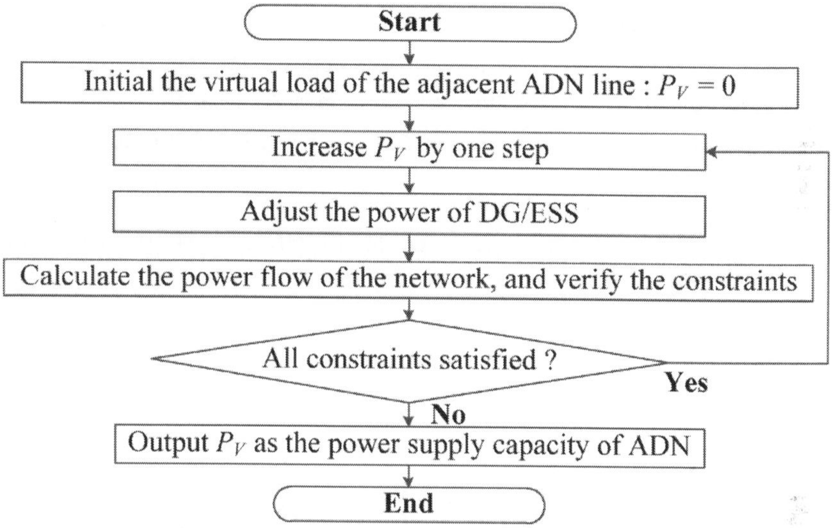

Figure 5. Block diagram of calculating the power supply capacity of an ADN line.

The process of the algorithm is shown in Figure 5. The iteration process of computing the power storage capacity of an ADN line is similar to that of power supply capacity, while a difference exists in the assumption of the adjacent line. The adjacent line is assumed to be a virtual energy source (P-Q bus), and the maximum power of the virtual energy source is the power storage capacity of the ADN line.

5. OPERATION OPTIMIZATION OF AN ADN CONSIDERING THE POWER SUPPLY RELIABILITY BASED ON THE POWER SUPPLY AND STORAGE CAPACITY

An operation optimization model is proposed in this section based on the aforementioned power supply and storage capacity of an LACR and ADN lines. With the active management of power controllable DGs and ESSs, the operation optimization of ADN brings economic benefits. The requirement for the network backup capacity will be reduced with the reduction of load peak-valley differences. The investment of upgrading the power facilities are postponed with the peak shaving, and the network losses are also reduced. The objective of the operation optimization in this paper is to maximize the economic benefits of ADN during 24 h. The LACR in the ADN line is operated as a whole, and the power of the LACR in 24 h is the variable to be optimized.

The economic benefit is calculated as Equation (15), in which the benefit consists of four parts: reduction of peak shaving cost, generation benefit of DGs, reduction of the network loss and the benefit of ESS charging/discharging at different electricity prices:

$$EB = E_{ps} \cdot p_{ps} + \sum_{t=1}^{24} E_{DG} \cdot p(t) + \sum_{t=1}^{24} \Delta E_{loss} \cdot p(t) + \sum_{t=1}^{24} E_{ess}^{D} \cdot p(t) \quad (15)$$

In Equation (15) E_{ps} is the energy of peak shaving of DG/ESS; and p_{ps} is the peak shaving cost of the grid; $E_{D,g}$ is the energy generated by power controllable DG; and $p(t)$ ΔE_{loss} is electricity price at time t; is the reduction of the network loss; E_{ess}^{D} is the energy discharged by ESS. A positive value of E_{ess}^{D} denotes discharged energy, while a negative value denotes charged energy.

The power supply reliability of an ADN line is considered in the optimization. When a fault occurs in an ADN line, the network is reconfigured, and the load of the faulty line is transferred to adjacent lines [20]. The maximum load transfer capacity of the ADN line is quantified by its power supply capacity according to the definition of the power supply capacity of the ADN line. Therefore, the constraint of power supply reliability is expressed as Equation (17a), in which $P_{su.N}^{\max}(t)$ is the power supply capacity of the ADN line at time t, and P_R is the outage load that needed to be transferred. The optimization model is shown by Equations (16) and (17):

$$\max(EB) = \max\left[E_{ps} \cdot p_{ps} + \sum_{t=1}^{24} E_{DG} \cdot p(t) + \sum_{t=1}^{24} \Delta E_{loss} \cdot p(t) + \sum_{t=1}^{24} E_{ess}^{D} \cdot p(t)\right] \tag{16}$$

$$\begin{cases} P_{su.N}^{\max}(t) \geq P_R(t) & \text{(a)} \\ -P_{st.k}^{\max}(t) \leq P_k(t) \leq P_{su.k}^{\max}(t) & \text{(b)} \\ V_{\min} \leq V_n(t) \leq V_{\max} & \text{(c)} \\ I_b(t) \leq I_{b.Rated} & \text{(d)} \\ 1 \leq t \leq 24 & \text{(e)} \end{cases} \tag{17}$$

The symbols in Equation (16) have the same meaning as those in Equation (15). Equation (17a) is the constraint of power supply reliability, and Equation (17b) is the power constraint of the kth LACR. $P_k(t)$ is the power of the kth LACR at time t, and the positive direction of is to inject into the ADN line. $P_{su.k}^{\max}$, $P_{st.k}^{\max}$ are the power supply/storage capacity of the k^{th}LACR, which are the upper and lower limitations of $P_k(t)$. Equations (17c,d) are the constraints of node voltage and branch current. Since the LACR is operated as a whole, the count of variables to be optimized is reduced. The power supply and storage capacity quantifies the power regulating range of the LACR and the maximum power supplied to adjacent lines, which reduces the count of constraints. Accordingly, the computation cost and complexity are reduced by applying the power supply and storage capacity in the optimization model.

6. CASE STUDY

The methodology presented in this paper is tested using a 10 kV active distribution network with three DGs, two ESSs and 22 loads, as shown in Figure 6. The network consists of two lines: line A and line B, and the two lines are connected together through one tie switch. Two LACRs exist in line A, and the rated maximum apparent power of the wires in the network is 4 MVA. The objective of line A's operation optimization is to maximize the economic benefits. The power supply reliability of line B is considered in the optimization of line A, and the load of line B will be transferred to line A if fault occurs at line B. The optimization model is given by Equations (16) and (17).

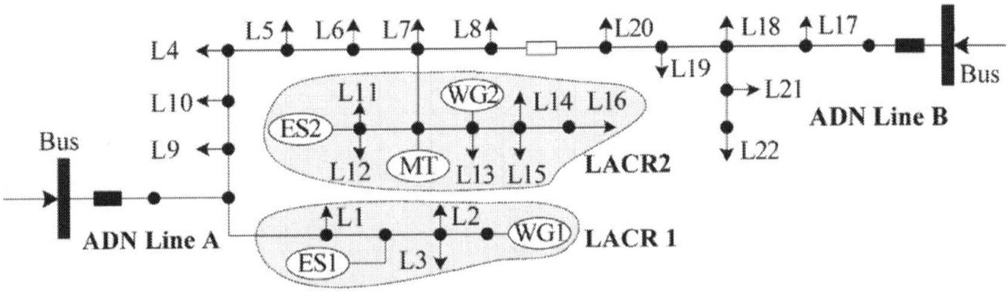

Figure 6. Topological connection of ADN.

The electrical specifications of the DGs/ESSs are shown in Table 1. ES1, ES2 and MT are dispatchable. The minimum operating power of MT is 10 kW, and the contract energy generated in 24 h is 2000 kWh. ES1 and ES2 are sodium-sulphur battery (NAS) energy storage systems. In practice, NAS batteries are limited to discharge less than 90% of their theoretical capacity to limit the corrosive properties of the material [21]. Correspondingly the upper and lower limitations of SOC in this case are set to 95% and 10%, respectively. The wind generation power is dependent upon wind speed, as shown inFigure 7. The electricity price is 0.68 CNY/kWh during the time [9,22], and is 0.33 CNY/kWh for the rest of the time. [17,19] period is peak time while [5,7] is the valley time. The peak shaving cost of the grid is 0.167 CNY/kWh [22].

Table 1. Electrical specifications of the DGs/ESSs.

Name	Type	Rated power	Energy capacity	Min operating power
ES1	NAS battery energy storage	300 kW	2400 kWh	0
ES2	NAS battery energy storage	200 kW	1600 kWh	0
MT	Micro turbine	280 kW	-	10 kW
WG1	Wind generation	500 kW	-	-
WG2	Wind generation	1000 kW	-	-

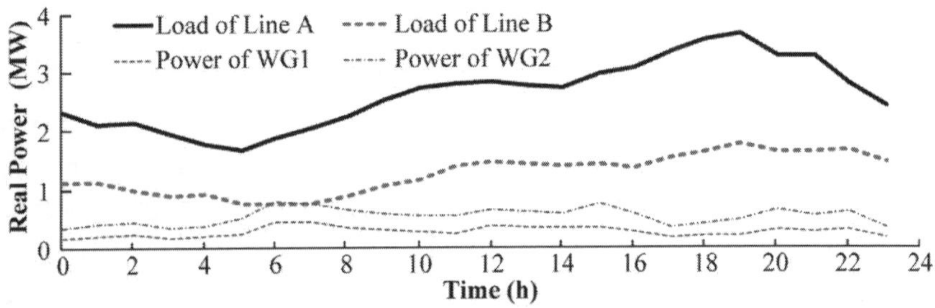

Figure 7. Load of ADN and power of wind generator.

The total load of ADN lines and the power of wind generations during 24 h are shown in Figure 7. The peak and valley load of line A are 3.65 MW and 1.65 MW, respectively, and the mean load is 2.61 MW. The mean square error of the load is 3.15×10^{11}. The power intermittence of wind generation increases the power fluctuation of line A. A genetic algorithm is implemented to solve the optimization. Figure 8 shows the optimum result of the charging/discharging power and SOC of ES1, ES2.

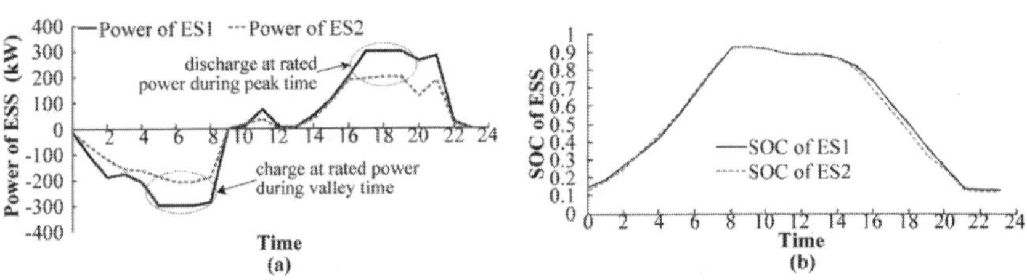

Figure 8. (**a**) Power of ES1 and ES2 after the optimization; (**b**) SOC of ES1 and ES2.

As shown in Figure 8, ES1 and ES2 charge during [0,8], and discharge during [10,23], maximizing the benefits of charging/discharging at different electricity prices. The discharging power reaches the rated ESS discharging power during the peak time, and the charging power reaches the rated charging power during the valley time, maximizing the energy of peak shaving.

Figure 9 shows the power of MT and the net load of line A after the optimization. As the contract energy of MT generated in 24 h is 2000 kWh, MT stops when the electricity price is 0.33 in order to maximize the benefit of electricity generation, and operates at its rated power during the peak time to maximize the energy of peak shaving. The peak-valley difference of the net load of line A is reduced from 2.33 MW to 1.05 MW after the optimization, as shown in Figure 9b.

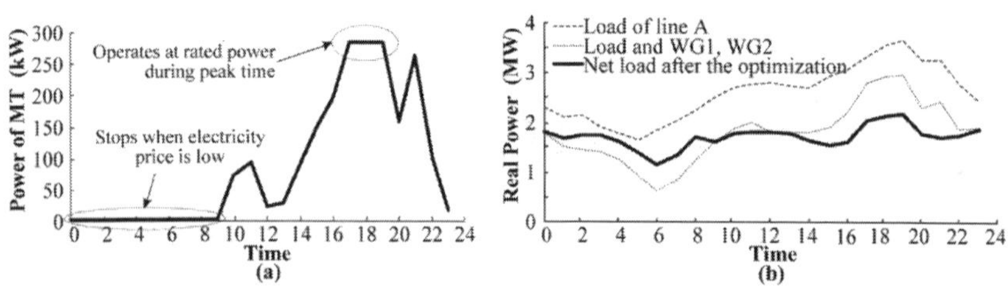

Figure 9. (**a**) Power of MT after the optimization; (**b**) Net load of line A after the optimization.

Without the operation optimization, ES1 and ES2 are assumed to charge/discharge at constant power considering the electricity price, and MT also operates at constant power during [7,23], as shown in Figure 10. The economic benefits of ADN with/without optimization are compared in Table 2.

Compared with the operation without optimization, the benefits of peak shaving and loss reduction increase by 74% and 22%. The generation benefit of MT and benefit of electricity price difference of ESS increase by only 7% and 14%, that is because the change of electricity price has already been considered in the operation of ESSs and MT without optimization, as shown in Figure 10. The total benefit is 3782 CNY after the optimization, while that is 3163 CNY before the optimization.

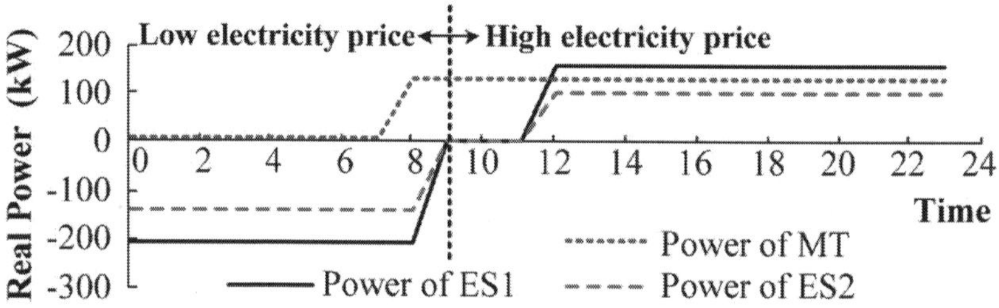

Figure 10. Operations of ES1, ES2 and MT without optimization.

Table 2. Comparison of economic benefits with/without the optimization.

Operation	Benefit of peak shaving		Generation benefit of MT	Benefit of loss reduction		Benefit of electricity price difference of ESS
Without optimization	2194 kWh	366 CNY	1272 CNY	594.7 kWh	527.8 CNY	998 CNY
With optimization	3812 kWh (174%)	637 CNY (174%)	1360 CNY (107%)	736.7 kWh (124%)	644.5 CNY (122%)	1141 CNY (114%)

The power supply & storage capacity of LACR1 and LACR2 in the optimization are shown in Figure 11. The optimized power of LACR1 and LACR2 are regulated within the power adjustable ranges determined by their power supply and storage capacity, and the Constraint (17b) is satisfied. The load is light in LACR1, and is less than the power of ES1 and WG1. Both of the power supply capacity and storage capacity of LACR1 are positive most of the time, as shown in Figure 11a, and LACR1 has energy storage system characteristics. LACR2 consists of ES2, MT and WG2, and the load is heavy in LACR2. The power supply capacity of LACR2 is negative most of the time, as shown in Figure 11b, and LACR2 has power controllable load characteristics.

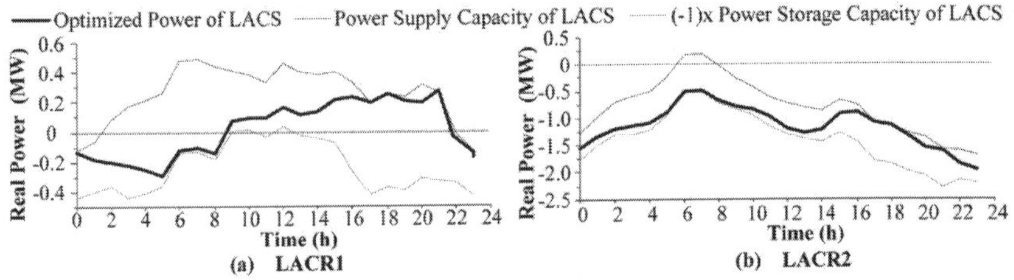

Figure 11. (**a**) Power supply and storage capacity and actual power of LACR1; (**b**) Power supply & storage capacity and actual power of LACR2.

Power supply reliability is considered in the optimization. The power supply capacity of line A before and after the optimization are calculated, as shown in Figure 12. The fault is assumed to occur at the outlet breaker of line B, and all the loads of line B need to be transferred to line A. During the time period [18,20] the loads of both line A and line B are heavy, and the power supply capacity of line A before the optimization cannot meet the requirement of the load transfer of line B. The operation optimization increases the power supply capacity of line A during the peak hours, and consequently the constraint of power supply reliability is satisfied.

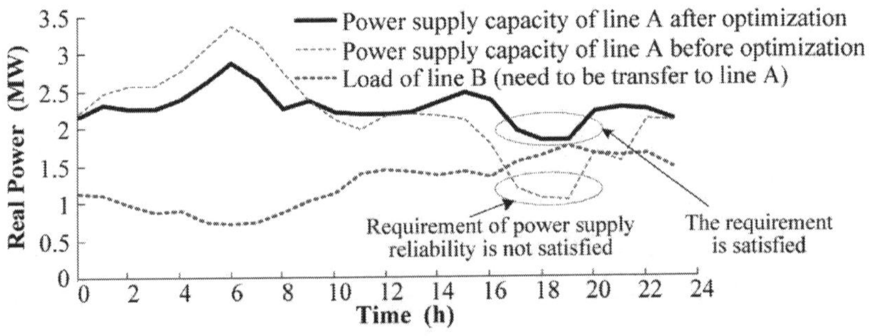

Figure 12. Power supply capacity of ADN Line A after optimization.

The computation cost of the optimization based on the power supply and storage capacity of the ADN is compared with that based on individual DGs/ESSs, as shown in Table 3. Both the optimization models are resolved by a genetic algorithm. Since the application of power supply and storage capacity reduces the number of variables and constraints in the optimization model, the number of iterations of the algorithm is reduced obviously (from 162 to 103). The power supply and storage capacity are calculated in the iteration of the algorithm, which increases the calculation time of the single cycle in the iteration. However, the total calculation time of the optimization based on power supply and storage capacity is reduced (from 215 s to 187 s).

Table 3. Computation cost comparison of the optimization in different models.

Optimization model	Algorithm	Number of iterations	Calculation time
Based on individual DG/ESS, without power supply &storage capacity	GA	162	215 s
Based on power supply & storage capacity	GA	103 (63.2%)	187 s (87.0%)

The calculation time is reduced by 13% with the application of power supply and storage capacity in the study case, in which there are only two ESSs and one DG to be optimized. As the number of DGs/ESSs increases, the benefit of applying the power supply and storage capacity will be more significant.

7. CONCLUSIONS

Grid-connected DGs and ESSs endow the distribution network with the ability of managing the electricity flows. Since the operation is subject to both DG's/ESS's electrical specifications and network constraints, the complexity of the operation optimization of an ADN is greatly increased with the number of DGs/ESSs. In this work, power supply and storage capacity indexes of a Local Autonomy Control Region are proposed to quantify the power regulating range of the LACR, considering both the power constraints of DGs/ESSs and network constraints. Similarly, the power supply and storage capacity indexes of an ADN line are presented to quantify the maximum power that can be exchanged between adjacent lines through tie switches. Afterwards, practical calculation methods are proposed for the indexes of LACR and ADN lines. Then the operation optimization of an ADN is proposed based on the indexes, which is proved effective by a study case.

The application of power supply and storage capacity indexes of a LACR can reduce the number of variables to be optimized in the optimization model. The number of constraints in the optimization model is also reduced since the power constraints of DGs/ESSs and network restrictions inside the LACR have already been considered in the model of the indexes. The power supply and storage capacity of an ADN line quantifies the maximum power that can be exchanged between ADN lines, which simplifies the power supply reliability constraint. In conclusion, both the computation cost and the complexity of the optimization model are reduced by applying the power supply and storage capacity.

ACKNOWLEDGMENTS

The authors would like to thank the support of The National High-tech R&D Program of China (863 Program with issue number: 2012AA050212).

CONFLICTS OF INTEREST

The authors declare no conflict of Interest.

REFERENCES

1 Chacra, F.A.; Bastard, P.; Fleury, G.; Clavreul, R. Impact of energy storage costs on economical performance in a distribution substation. *IEEE Trans. Power Syst.* **2005**, *2*, 684–691.

2 Schoenung, S.; Burns, C. Utility energy storage applications studies. *IEEE Trans. Power Syst.* **1996**, *3*, 658–665.

3 Arani, M.F.M.; El-Saadany, E.F. Implementing virtual inertia in DFIG-based wind power generation. *IEEE Trans. Power Syst.* **2013**, *2*, 1373–1384.

4 Chen, S.-S.; Wang, L.; Lee, W.-J.; Chen, Z. Power flow control and damping enhancement of a large wind farm using a superconducting magnetic energy storage unit. *IET Renew. Power Gener.* **2009**, *1*, 23–28.

5 Lu, M.-S.; Chang, C.-L.; Lee, W.-J.; Wang, L. Combining the wind power generation system with energy storage equipment. *IEEE Trans. Ind. Appl.* **2009**, *6*, 2109–2115.

6 Li, B.; Shen, H.; Tang, Y.; Wang, H. Impacts of energy storage capacity configuration of HPWS to active power characteristics and its relevant indices. *Power Syst. Technol.* **2011**, *4*, 123–128.

7 Vazquez, S.; Lukic, S.M.; Galvan, E.; Franquelo, L.G.; Carrasco, J.M. Energy storage systems for transport and grid applications. *IEEE Trans. Ind. Electron.* **2010**, *12*, 3881–3895.

8 Bao, G.; Lu, C.; Yuan, Z.; Lu, Z. Load shift real-time optimization strategy of battery energy storage system based on dynamic programming. *Autom. Electr. Power Syst.* **2012**, *12*, 11–16.

9 Yu, P.; Zhou, W.; Sun, H.; Guo, L.; Sun, F.; Sui, Y. Hybrid energy storage system and control system design for wind power balancing. *Proc. CSEE* **2011**, *17*, 127–133.

10 Choi, M.-E.; Kim, S.-W.; Seo, S.-W. Energy management optimization in a battery/supercapacitor hybrid energy storage system. *IEEE Trans. Smart Grid* **2012**, *1*, 463–472.

11 Cie, Q.; Wang, Y.; Kim, Y.; Pedram, M.; Chang, N. Charge allocation in hybrid electrical energy storage systems. *IEEE Trans. Comput. Aided Des. Integr. Circuits Syst.* **2013**, *7*, 1003–1016.

12 Pilo, F.; Pisano, G.; Soma, G.G. Optimal coordination of energy resources with a two-stage online active management.*IEEE Trans. Ind. Electr.* **2011**, *10*, 4526–4537.

13 Currie, R.A.F.; Ault, G.W.; Foote, C.E.T.; McDonald, J.R. Active power-flow management utilising operating margins for the increased connection of distributed generation. *IET Gener. Transm. Distrib.* **2007**, *1*, 197–202.

14 Jing, Y.; Bai, H.; Zhang, J. Multi-objective optimization design and operation strategy analysis of a solar combined cooling heating and power system. *Proc. CSEE* **2012**, *20*, 82–87.

15 Hashemi, R.A. Developed offline model for optimal operation of combined heating and cooling and power systems.*IEEE Trans. Power Syst.* **2009**, *1*, 222–229.

16 D'adamo, C.; Jupe, S.; Abbey, C. Global Survey on Planning and Operation of Active Distribution Networks-Update of CIGRE C6.11 Working Group Activities. In Proceedings of the 20th International Conference and Exhibition on Electricity Distribution: Part 1, Prague, Czech, 8–11 June 2009; pp. 1–4.

17 Farag, H.E.; El-Saadany, E.F.; El Chaar, L. A Multilayer Control Framework for Distribution Systems with High DG Penetration. In In Proceedings of the 2011 International Conference on Innovations in Information Technology (IIT), Abu Dhabi, United Arab Emirates, 25–27 April 2011; pp. 94–99.

18 You, Y.; Liu, D.; Yu, W.; Chen, F.; Pan, F. Technology and its trends of active distribution network. *Autom. Electr. Power Syst.* **2012**, *18*, 10–16.

19 Zhang, J.; Cheng, H.; Wang, C.; Xia, Y.; Shen, X.; Yu, J. Quantitive Assessment of Active Management of Distribution Network with Distributed Generation. In Proceedings of the Third International Conference on Electric Utility Deregulation and Restructuring and Power Technologies, 2008, (DRPT 2008), Nanjing, China, 6–9 April 2008; pp. 2519–2524.

20 Liao, H.; Liu, D.; Huang, Y.; Zhang, Y. Load transfer capability analysis considering interconnection of distributed generation and energy storage system. *Int. Trans. Electr. Energy Syst.* **2012**.

21 Sarasua, A.E.; Molina, M.G.; Pontoriero, D.E.; Mercado, P.E. Modelling of NAS Energy Storage System for Power System Applications. In Proceedings of the 2010 IEEE/PES Transmission and Distribution Conference and Exposition: Latin America (T&D-LA), Sao Paulo, Brazil, 8–10 November 2010; pp. 555–560.

22 Chen, Q.-X.; Kang, C.-Q.; Xia, Q. Operation mechanism and peak-load shaving effects of carbon-capture power plant.*Proc. CSEE* **2010**, *30*, 22–28.

CHAPTER 8

Integrated Electrical and Thermal Grid Facility - Testing of Future Microgrid Technologies

Sundar Raj Thangavelu [†], **Inam Ullah Nutkani** [†], **Chia Meng Hwee** [†], **Aung Myat** [†] **and Ashwin Khambadkone** [†,*]

Experimental Power Grid Centre, Institute of Chemical and Engineering Sciences, 3 Pesek Road, Singapore 627590, Singapore

ABSTRACT

This paper describes the Experimental Power Grid Centre (EPGC) microgrid test facility, which was developed to enable research, development and testing for a wide range of distributed generation and microgrid technologies. The EPGC microgrid facility comprises a integrated electrical and thermal grid with a flexible and configurable architecture, and includes various distributed energy resources and emulators, such as generators, renewable, energy storage technologies and programmable load banks. The integrated thermal grid provides an opportunity to harness waste heat produced by the generators for combined heat, power and cooling applications, and support research in optimization of combined electrical-thermal systems. Several case studies are presented to demonstrate the testing of different control and operation strategies for storage systems in grid-connected and islanded microgrids. One of the case studies also demonstrates an integrated thermal grid to convert waste heat to useful energy, which thus far resulted in a higher combined energy efficiency. Experiment results confirm that the facility enables testing and evaluation of grid technologies and practical problems that may not be apparent in a computer simulated environment.

KEYWORDS

microgrid test facility; distributed generation; integrated electrical and thermal grids; flexible and configurable architecture; energy efficiency

1. INTRODUCTION

Distributed Generation (DG) is driving a paradigm shift in the traditional power system. The common approach for using DG is by forming a microgrid—A controllable entity consists of interconnected distributed energy resources (renewable, conventional combustion engine based generators and/or storage systems) and loads (electrical and thermal) that can balance its supply and demand locally, and can operate in islanded and/or grid-connected mode—which has emerged as a more viable solution to increase the power system efficiency, reliability and security [1,2]. The microgrids, being close to the load, increase overall efficiency of the system by saving towards transmission losses and waste heat recovery for heating or cooling applications.

Over the past decade, the microgrids have been demonstrated successfully for several technology configurations and applications, particularly in North America, Europe, and Asia [3,4,5]. According to a research report [6], the current (2Q2015) global installed microgrid capacity is estimated approximately 12 GW, out of which 47% in Asia, 44% in North America and 9% in Europe. The widespread adoption of microgrids, however, is subject to overcoming of economic and institutional barriers. These barriers can be overcome by the reduction in the technology cost, offering proper incentives and utility tariff, and formation of the regulatory framework, including defining of grid interconnections and performance requirements and microgrid enabling market mechanism [7,8].

Considerable effort has already been made in defining the performance and interconnection requirements for the DG and microgrids. So far, the key development is the guideline standard for the DG, IEEE 1547 Series 1–3: Interconnecting Distributed Resources with Electric Power Systems [9] and for the microgrid, IEEE 1547.4: Guide for Design, Operation, and Integration of Distributed Resource Island Systems with Electric Power Systems [10]. These standards together with local grid codes have been widely adopted in the US. Similarly, the DG interconnection standards have been developed and adopted by the other countries, worth mentioning are Europe [11,12,13] and Japan [14,15,16]. The general outlines of the interconnection standards and local grid codes adopted by different countries are more or less similar; however, the specific performance parameters and test procedures have been tailored according to the local conditions and thus vary from country to country. The most important performance tests covered under these standards and codes are related to power quality, protection, fault ride through, grid synchronization, anti-islanding and reconnection. The future grid-codes are expected to include new tests and requirements such as active power curtailment, reactive power and voltage support, and low voltage (LV) and high voltage (HV) ride through capabilities [17,18]. In addition, future microgrid in industries and urban regions aims to improve energy efficiency to remain economical, one of the likely option is integrated thermal grid to recover waste heat from distributed generation into useful energy [19,20,21].

Each of the aforementioned performance tests for the DG and microgrids requires a flexible and specially designed experimentation facility. For the widespread adoption of microgrids, there is a need for such facilities that allow microgrid developers and service providers to test and validate their solutions against the local standards and codes before real deployment.

The Experimental Power Grid Centre (EPGC) has a unique microgrid test facility, rated at up to 1 MW, specially designed to perform a wide range of DG and microgrid tests, including those listed in the existing and being discussed in the upcoming standard and grid codes. Against a backdrop of rapid urbanization in the Asia Pacific region and national initiatives to make Singapore a Smart Nation and improve its status as a role model for sustainability and liveability, the facility was developed in line with EPGC's mission to work with industries, both local and international, research institutions and government organizations driving technological development in the field of new and advanced power systems.

This paper presents a design overview of the EPGC microgrid facility with integrated electrical and thermal grids, which allows test-bedding and research, development and demonstration (RD & D) of a variety of DG and microgrid technologies. These technologies and solutions can be applied to a growing urban landscape where, for example, microgrids can be a solution for integrating more renewable and conventional technologies for remote applications of which an increasing number exist in the surrounding region. The paper further discusses a few cases of microgrid testing with the support of experimental results. The test cases elaborated in the subsequent sections of the paper are related to dynamic control of storage systems, microgrid islanding and reconnection, and microgrid frequency and voltage regulation using storage systems. A case study for heat recovery from the diesel generator exhaust is also presented, where the recovered heat energy is used to support air-conditioning in a tropical climate.

2. ARCHITECTURE OF THE INTEGRATED MICROGRID FACILITY AT EXPERIMENTAL POWER GRID CENTRE (EPGC)

This section will highlight the features of EPGC facility comprising electrical grid and integrated thermal grids.

2.1. Electrical Grid Architectures and Functionalities

The EPGC microgrid topology is shown in the Figure 1. This facility has 400 V 4 buses power network, each bus consists of 3 phase 4 wires. Buses can be operated independently or coupled to form various topologies ranging from radial, series, loop with the ability to vary bus impedances mimicking the distance between the buses. Figure 2 shows the picture of the actual bus panels.

Table 1 and Table 2 shows the DG in EPGC microgrid comprising generators, emulators, photovolatic arrays, storage technologies and load banks [22,23]. At EPGC, three types of emulators are used for various experiments. Emulators are devices that mimic the behaviour of an actual machine; for example, a wind turbine emulator (WTE) allows us to input a wind velocity profile and the WTE will create the electrical power injection characteristics of a wind turbine generator connected to the experimental grid. Similarly for a photovoltaic (PV) emulator, it is able to inject DC power similar in character to a real PV array in various states of irradiance. The Power Grid emulator allows the creation of varying grid conditions to facilitate the study of the grid behaviour during different grid conditions without being connected to the power grid. Emulators provide the freedom from the geographical constraints of being located in Singapore. Wind profiles and irradiance profiles can be used to virtually transport the grid to various locations in the world. Furthermore, it provides repeatability when conducting experiments. In addition to the experimental grid, the researchers at EPGC also utilise various simulation and modelling platforms. Researchers can analyse various phenomenon in simulation before taking the next step in validating their models in the experimental grid.

2.2. Thermal Grid Architectures and Functionalities

The potential opportunity to improve the overall energy efficiency of the microgrid is in implementing the thermal grid to recover waste heat from the exhaust gas and convert into various useful forms of energy [20,24]. Figure 3 shows the thermal grid architecture which includes critical systems such as waste heat recovery system, absorption/adsorption chiller and thermal storage systems responsible for recovering waste energy and generating useful forms of thermal energy.

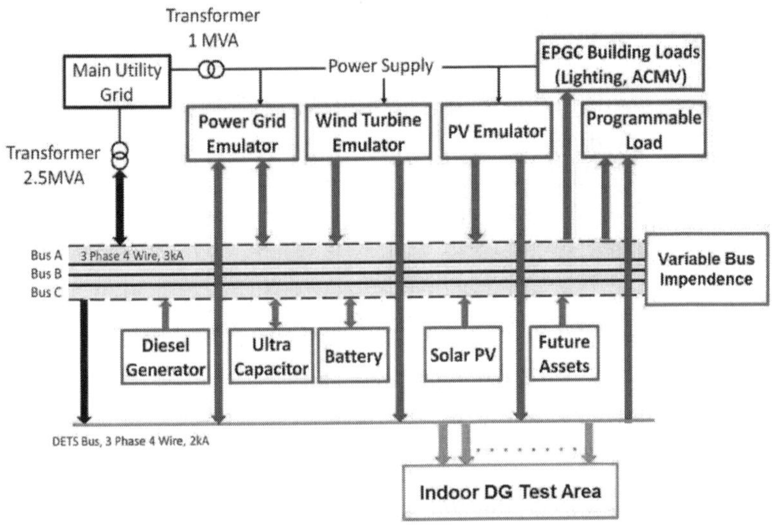

Figure 1. Schematic diagram of EPGC's microgrid facility.

Figure 2. 3000 A, 400 V, 3 phase 4 wire Buses Panel.

Table 1. Experimental grid assets.

Diesel Generators	Rating	Storage Technologies	Rating
DG1	250 kVA	Lithium Ion	2× 20 kWh
DG2	125 kVA	Lead Acid	110 kWh
DG3	50 kVA	Ultra-capacitors (Electric Double Layer Capacitor)	90 kW for 30 s
			60 kW for 30 s
Photovoltaic Arrays	**Rating**	**Emulators**	**Rating**
Monocrystalline	20 kWp	Power Grid Emulator	90 kW
Polycrystalline	20 kWp	Wind Turbine Emulator	10 kW
Amorphous Thin Film	20 kWp	PV Array Emulator	30 kW

Table 2. Programmable load banks.

Rated Capacity		
Resistive (kW)	**Inductive (kVAR)**	**Capacitive (kVAR)**
627	500	380

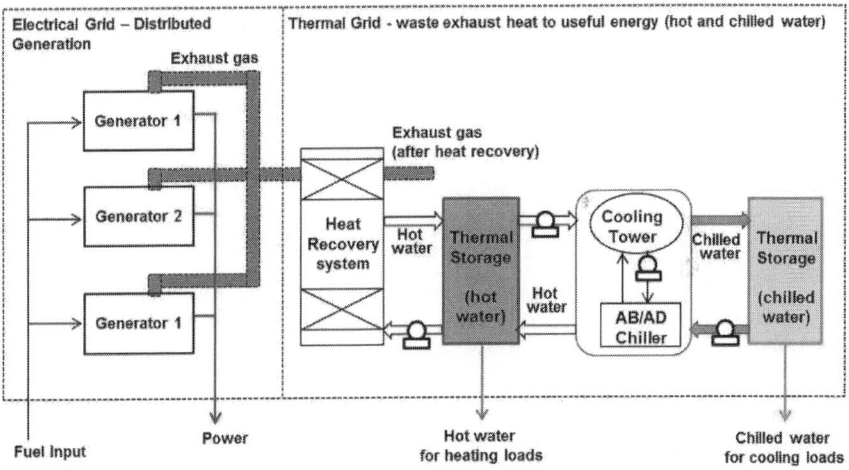

Figure 3. Thermal grid recovering waste heat into useful thermal energy.

2.2.1. Waste Heat Recovery System to Support Hot Thermal Loads

The potential waste heat available in EPGC microgrid are from three generators (50, 125, 250 kVA), load banks and solar-PV. A thermal grid is implemented and currently powered by the exhaust gas from generators to investigate the potential energy efficiency opportunities. Waste heat recovery system (WHRS) recovers waste heat in the form of steam or hot water depending on the heat potential of exhaust gas and the type of thermal loads [25]. A shell and tube heat exchanger is designed as WHRS based on the generator size and, the thermal and chemical properties of exhaust gas [26]. The designed WHRS recovers 55%–80% of heat from the exhaust gas, whereas the rest is rejected to the atmosphere considering the thermodynamic and design limits.

By implementing the WHRS system, the overall energy efficiency is increased from ηDER to ηDER+WHRS as shown in Equation (1)–Equation (3),

$$\eta_{DER+WHRS} = \frac{\textit{useful energy}}{\textit{fuel energy content}} = \frac{E_{power} + E_{hot\ water}}{E_{fuel}} \quad (1)$$

$$\eta_{DER+WHRS} = \frac{E_{power}}{E_{fuel}} + \frac{E_{fuel} - E_{power}}{E_{fuel}} \frac{E_{hot\ water}}{E_{fuel} - E_{power}} \quad (2)$$

$$\eta_{DER+WHRS} = \eta_{DER} + (1 - \eta_{DER})\eta_{WHRS} \quad (3)$$

where, the combined energy efficiency (ηDER+WHRS) of diesel generator and waste heat recovery system is the ratio of useful energy (electrical energy Epower and thermal energy Ehotwater) to the energy content of the fuels (Efuel). ηDER is the electrical efficiency of the diesel generator, which typically varies between 20% to 35%, while operating at loads between 50% to 100% (full load condition). ηWHRS is the efficiency of a waste heat recovery system which recovers heat energy from the exhaust gas to useful thermal energy such as hot water up to 90 ∘C. The combined efficiency (ηDER+WHRS) ranges between 60%–74%. Thermal grid with WHRS can support heating loads such as steam and hot water. The real benefits come when the heating loads locates near the WHRS otherwise the pumping cost need to be considered. When generated steam or hot water is higher than the heating loads, either thermal storage is a preferable option to store excess energy for later use or convert to other forms of useful energy such as chilled water depending on the cooling loads.

2.2.2. Waste Heat Driven Absorption Chiller to Support Cold Thermal Loads

The waste heat recovered in the form of steam or hot water is the main driver in the absorption system to produce chilled water. The refrigerant absorbs heat from chilled water (returning from air conditioning system) at a low temperature and pressure during evaporation, and release heat to the cooling water (to cooling tower) at a high temperature and pressure heat during condensation. The vaporised refrigerant (after evaporation) is absorbed into the absorbent at low pressure, followed by heating at high temperature to vaporize and restore its original concentration. Water and Lithium Bromide are the refrigerant and absorbent used in the absorption chiller. The recovered waste heat is used to vaporize the refrigerant for re-use. The performance of absorption chiller is represented as coefficient of performance (COP) or ηCH that is the ratio of cooling capacity generated in the chilled water to the heat input *i.e.*, heat supplied to the chiller as hot water. The overall energy efficiency of microgrid with integrated thermal grid is represented by ηDER+WHRS+CH (Equation (4)),

$$\eta_{DER+WHRS+CH} = \eta_{DER} + (1 - \eta_{DER})\eta_{WHRS}\eta_{CH} \tag{4}$$

ηDER+WHRS+CH is the combined energy efficiency of diesel generator, waste heat recovery system and chiller describing the ratio of useful energy to the energy content of the fuels. The electrical efficiency of diesel generators (less than 250 KVA) at rated condition is 35%, implementing the WHRS increases the combined energy efficiency to 74%. In case of chilled water loads, the combined efficiency is 60%–85% depending on the absorption chiller type [27]. The COP or ηCH of single effect absorption chiller is 0.65-0.8 (for low temperature hot water driven), 0.7–0.75 for medium temperature hot water driven, and 0.7–0.75 for steam driven chillers. The COP or ηCH of double effect absorption chiller is 1.4–1.45 (for high temperature hot water driven) and 1.38–1.43 for steam driven chiller whereas the triple effect absorption chiller (hot water or steam) achievesηCH to 1.8 [28,29]. This facility is installed with a 10 refrigeration ton of single effect absorption chiller driven by low temperature hot water at 90 ∘C to generate chilled water at 7 ∘C.

2.2.3. Hot and Cold Thermal Storage System

Thermal storage is cheaper than electrical storage and the energy cost is dominated by the operation cost unlike the capital cost and lifecycles of electrical storage system. The key purpose of thermal storage is to act as a buffer during the thermal supply and demand mismatch [24]. The thermal capacity of the storage system is effectively utilized by charging excess energy to use for peak shaving and peak shifting operations. Optimizing the storage temperature and advancing to phase-change-material (PCM) based storage are the possible options to increase storage capacity without changing the footprint area. Figure 4 shows the configuration of four cubic meter (2 × 2 × 1 m3) hot and cold thermal storage designed based on the WHRS, chiller system and air conditioning loads. The designed thermal storage can support the hot water and chilled water supply up to 1 h when DGs, WHRS and chiller not in operation.

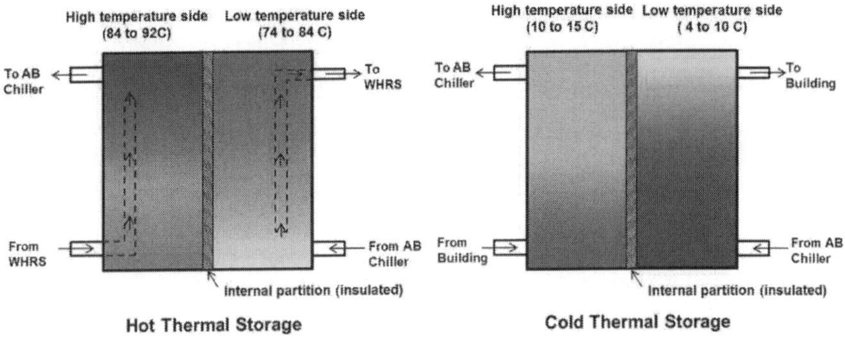

Figure 4. Schematic configuration of hot and cold thermal storage.

3. EXPERIMENTAL CASE STUDIES

This section discuss the experimental case studies investigating the control aspects of different microgrid topologies and the potential of optimized control to improve the overall energy efficiency of microgrid.

3.1. Coordinated Control of Hybrid Storage System in Grid-Connected Microgrid

By definition, microgrid is a single controllable entity that can balance its supply and demand locally without causing any disturbance to the utility grid. For the planning and designing of such microgrids and its control, EPGC experimental facility can be used to test and evaluate the performance of different microgrids configurations and control algorithms. Test case in this section demonstrates dynamic coordinated control of two different types of storage systems operating in the microgrid. During the transients, the considered microgrid can balance its supply and demand without causing any disturbance to the utility grid. As shown in Figure 5a, the microgrid comprises Diesel Generator (DG2), Electric Double-Layer Capacitor (EDLC1), and Programmable Load Bank (PLB2), where power flow control at Point of Common Coupling (PCC) is set equal to zero.

The experimental results for two different cases of load change in the microgrid are shown in Figure 6. The control mode for EDLC1 is set to 'local follow' as shown in Figure 5b and LeadAcid is set to "local follow" and "with EDLC" in order to compensate the rapid change in power at PCC, as shown in Figure 5b,c. This coordinated control behavior can be witnessed from the experimental results shown in Figure 6, where EDLC1 promptly responds to the load change at 25:24 s by charging 50 kW power for a short duration. At the same time, LeadAcid storage also responds to the load change, but at the slower rate than EDLC1, which is for obvious reasons. This coordinated dynamic control of two different storage systems allows DG2 to ramp down according to its design characteristics without causing any sudden power fluctuation at the PCC, which is the main objective of this control strategy.

3.2. Microgrid Islanding and Reconnection with Utility Grid

Another desired feature of the practical microgrid is to operate in islanded and grid-connected mode. According to standard 1547, the grid-connected DGs and microgrid shall transition from grid-connected to islanded mode during the grid disturbance and re-connect back once the grid is normal, within specified time duration. This transition shall be smooth such that the islanded microgrid shall continue to balance its supply and demand. EPGC facility is designed to perform anti-islanding tests for the DG and microgrids according to the standard IEEE 1547. This case study demonstrates an example of smooth transition from grid-connected mode to islanded-mode and vice versa, for the configured microgrid shown inFigure 7. The experimental results for the microgrid islanding and reconnection are shown in Figure 8. At operating point 1, the microgrid transitions into islanded mode while it is importing contracted power of 40 kW from the utility grid named as "Singapore Power PowerGrid (SPPG)". The power imbalance due to islanding is compensated by DG2 followed by LeadAcid which is operating in local follow mode.

At operating point 2, the microgrid is commanded to re-connect and synchronize with the utility grid (*i.e.*, SPPG) to continue importing the contracted power of 40 kW. As shown from the experimental results in Figure 8, the transition from islanded-mode to grid-connected mode is very smooth, where LeadAcid plays an important role to maintain the steady ramp-rate at PCC point whereby allowing DG2 to ramp down according to its characteristics.

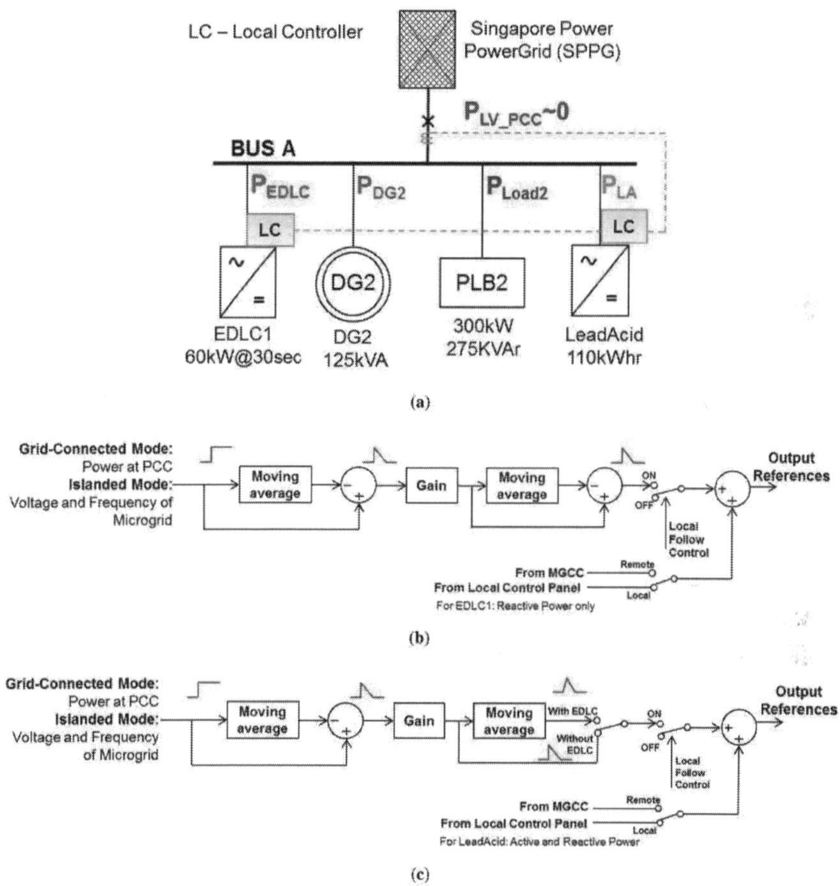

Figure 5. Microgrid configuration and control schemes during grid-connected and islanded mode operation. (**a**) Configured microgrid for storage control test; (**b**) Control scheme for EDLC1; (**c**) Control scheme for LeadAcid storage.

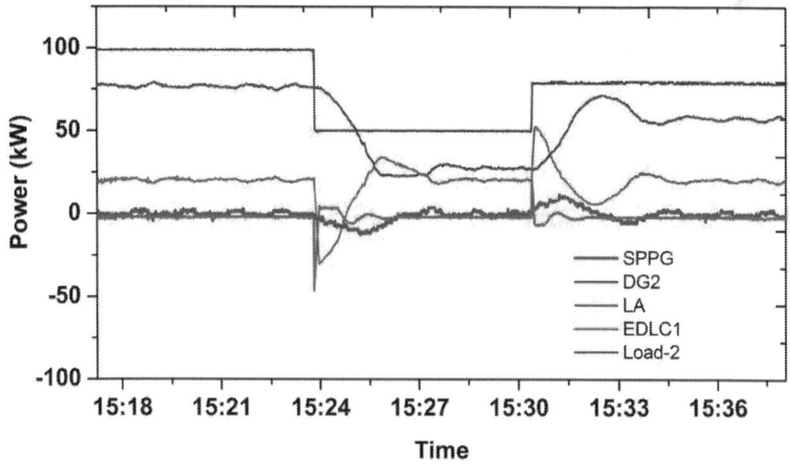

Figure 6. Experimental results showing EDLC1 and LeadAcid coordinated control to maintain PCC power 0 during sudden load change in microgrid.

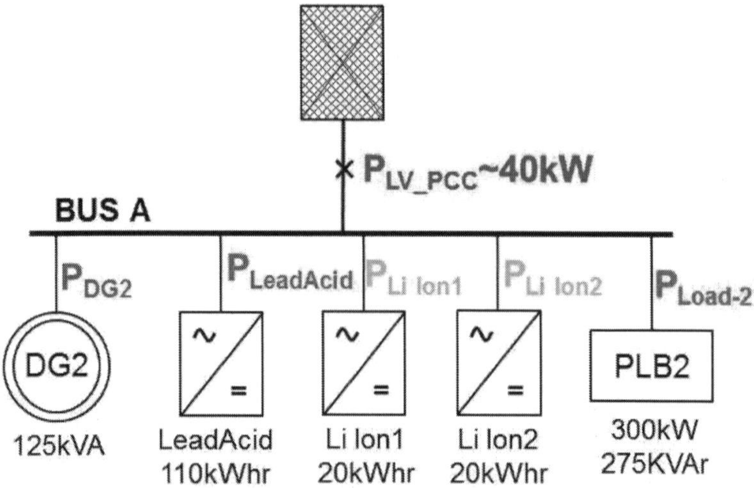

Figure 7. Configured microgrid for islanding and reconnection test.

3.3. Storage Application for Frequency and Voltage Regulation in Microgrid

In this case, the islanded microgrid shown in Figure 9 is tested with and without energy storage. The fast response energy storage system plays important role in maintaining the power quality in the low inertia microgrids, which has been demonstrated through this test case. The test results demonstrate that the frequency and voltage of low inertia microgrid can be regulated with proper control of storage system. The control scheme for LeadAcid and EDLC1 is shown in Figure 5b,c. With the common objective to improve the voltage and frequency variation during the load change (up and down), three different scenarios are tested as briefly described below.

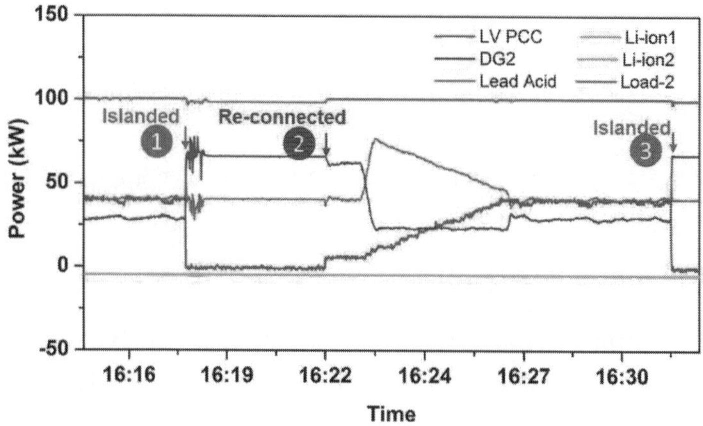

Figure 8. Experimental results showing microgrid islanding and reconnection with utility grid while importing 40 kW during grid-connected mode.

Scenario 1: LeadAcid only responds to the frequency and voltage change according to the control scheme shown in Figure 5c.

Scenario 2: EDLC1 only responds to the frequency and voltage change according to the control scheme shown in Figure 5b.

Scenario 3: Both EDLC1 and LeadAcid respond to the frequency and voltage change, as shown in Figure 5b,c. In this case, EDLC and LeadAcid response is coordinated, where ELDC lead by providing the faster response and LeadAcid provide slow response while "without EDLC" mode is activated, as shown in Figure 5c.

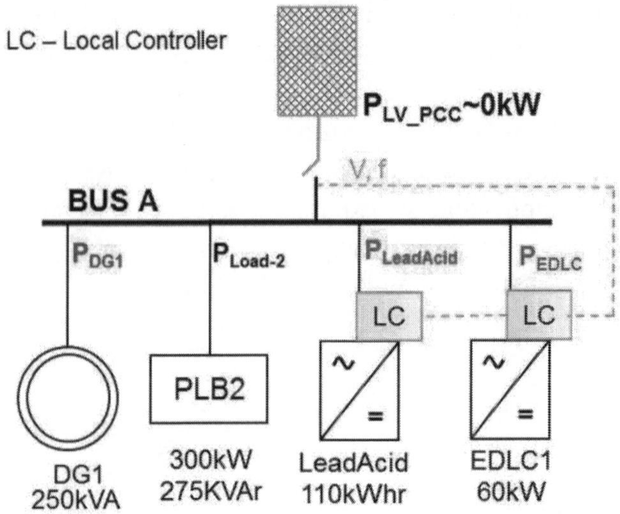

Figure 9. Configured islanded microgrid for storage application test.

The experimental results for the microgrid frequency and voltage variation during the step load change (up and down) are shown from Figure 10, Figure 11, Figure 12 and Figure 13 and summarized in Table 3.

As shown in Figure 10, the microgrid frequency varies from 50 to 45.77 Hz (8.5%) and voltage varies from 410 to 347 V (15%) when load changes from 50 kW, 0 kVAr to 150 kW, 50 kVAr during DG2 operation only. The same load change test is then repeated after activating the LeadAcid (LA) storage local follow control, which shows significant improvement in the microgrid frequency and voltage. With DG2 and LA storage, the microgrid frequency varies from 50 to 48.97 Hz (2%) and voltage varies from 410 to 39 5V (3.5%), which is significantly better than without storage. Similarly, the experimental results for the load step-down test also demonstrate the frequency and voltage regulation in the microgrid when storage system local follow control is activated.

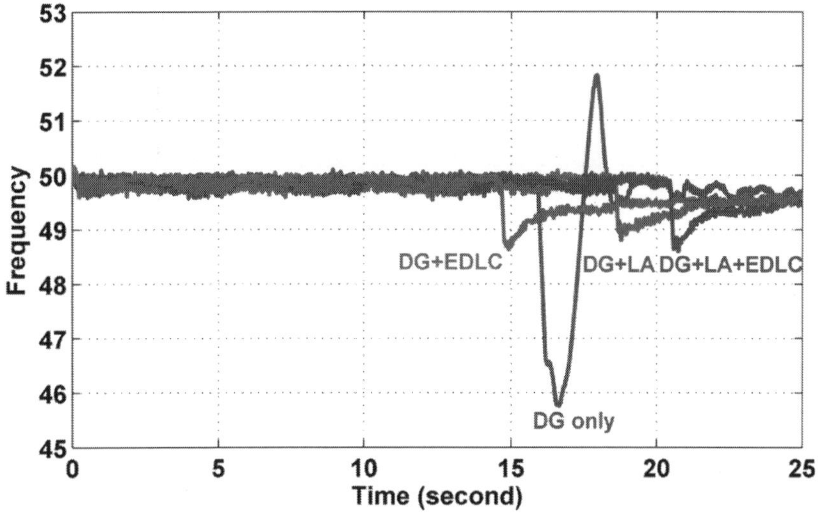

Figure 10. Experimental result showing microgrid frequency with and without storage during load step up from 50 kW, 0 kVAr to 150 kW, 50 kVAr.

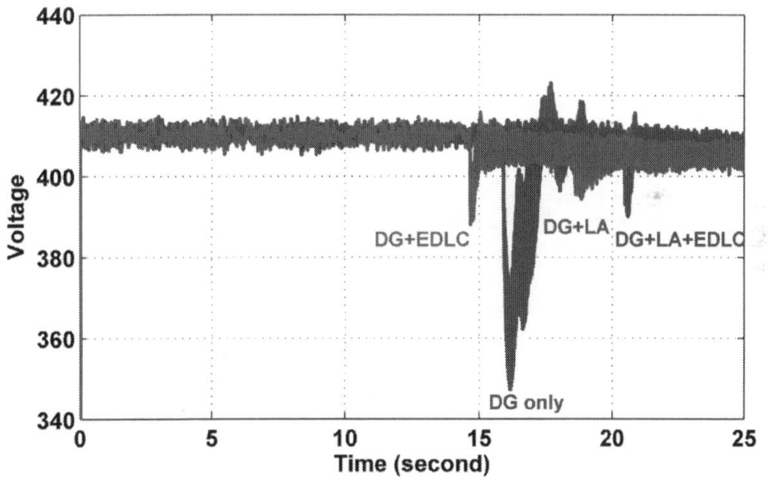

Figure 11. Experimental result showing microgrid voltage with and without storage during load step up from 50 kW, 0 kVAr to 150 kW, 50 kVAr.

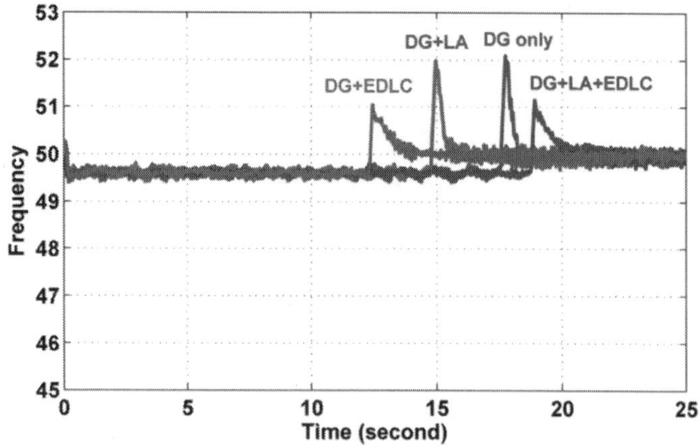

Figure 12. Experimental result showing microgrid frequency with and without storage during load step down from 150 kW, 50 kVAr to 50 kW, 0 kVAr.

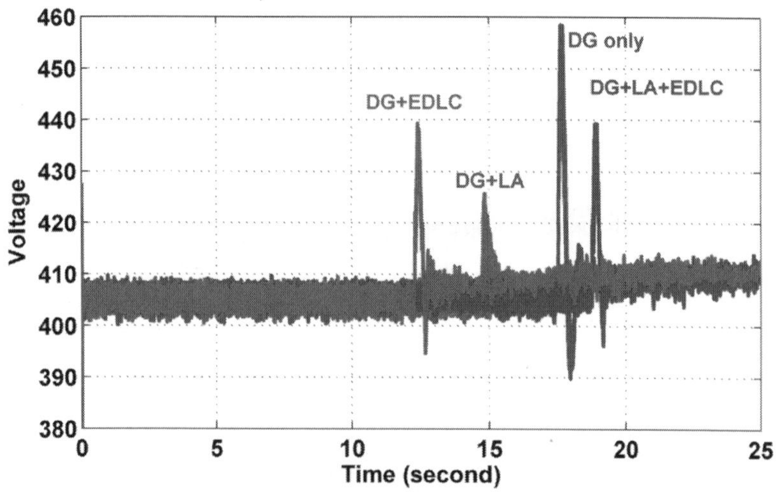

Figure 13. Experimental result showing microgrid voltage with and without storage during load step down from 150 kW, 50 kVAr to 50 kW, 0 kVAr.

Table 3. Summary of frequency and voltage variation with and without storage during step load change.

Load Test	Parameter	DG only	DG1+LeadAcid	DG1+LA+EDLC1	DG1+EDLC1
Step Up	Frequency	−4.23 to 1.82 Hz	−1.03 Hz	−1.19 Hz	−1.26 Hz
Step Down	Frequency	2.09 Hz	1.94 Hz	1.03 Hz	1.04 Hz
Step Up	Voltage	13.1 to −62.6 V	−14.5 V	5.7 to −20.0 V	5.9 to −23.1 V
Step Down	Voltage	56.1 to −12.2 V	22 V	34.3 to −7.4 V	34.7 to −9.6 V

3.4. Control of Distributed Generators

In simple microgrids, generation sources are often located close to each other. However, as microgrids grow and proliferate around the world, the generation sources within the microgrid can become more distributed. For example, renewable generation sources such as wind sources can be located at one side of the microgrid while main generators are located at the other side. In addition, microgrids can eventually connect with one another for greater reliability and resilience where generation from one microgrid can supply to another in times of need. However, such distributed generation or interconnection between microgrids may cause certain issues to arise especially when the generation sources have different control strategies. In the following experiment, a simple case of 2 generators connected via a line impedance has been used to showcase some problems that may occur in such situations. Even though additional sources and loads could have been easily added to this case, it has been simplified to highlight potential problems in distributed generation systems.

This case study experimentally investigates the workability of different control strategies for distributed generators. In the experimental setup, a 200 kW diesel generator (DG1) is run with a 100 kW diesel generator (DG2) separated by a line impedance of 20 + j33 mΩ as shown in Figure 14. This line impedance is almost equivalent to 400m line length. This setup emulates a possible distributed generation scenario where sources connect to each other via short lengths of cable or overhead line. An example of such scenarios can occur in college campus microgrids where generation and load centers are distributed across a geographical area.

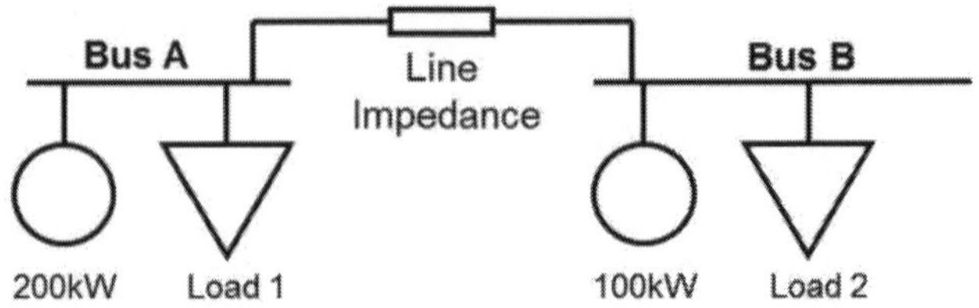

Figure 14. Setup of experiment.

In this experiment, two different control strategies have been examined. In the first control scheme, both DGs run in Power-frequency and Reactive Power-Voltage droop modes. The control block diagrams are shown in Figure 15. Loads 1 and 2 were then changed according to Table 4. The active power, reactive power, voltages and frequency were then recorded.

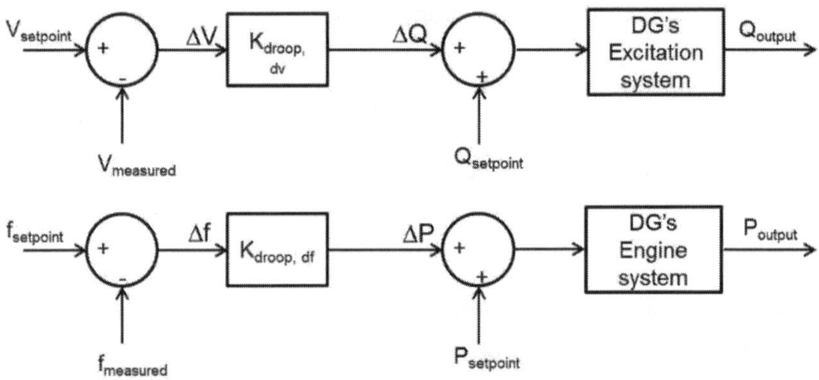

Figure 15. Droop control block diagrams for both DG1 and DG2.

Figure 16, Figure 17 and Figure 18 show that the active power varies according to the frequency output while the reactive power varies according to the voltages on the respective buses. The voltages on the buses vary according to the power transfer from Bus A to Bus B and vice versa. It can be noted that there is slight frequency oscillation at a loading of 150 kW on bus A. This can be due to the interaction between the two DG controllers. Another problem that can be noted is due to the voltage droop characteristics, 1 DG would supply reactive power while the other consumes it. This increases the amount of losses in the line due to a higher current flow through the line.

Table 4. Load profile of loads 1 and 2.

Number	Time (s)	Load 1 (kW)	Load 2 (kW)
1	0	50	150
2	192	50	50
3	560	150	50
4	826	150	0

In the second set of experiments, the control strategy has been changed to a Master-slave control scheme in which DG1 maintains the microgrid voltage and frequency to 415 V and 50 Hz respectively. DG2 controls the active power output to a setpoint of 80 kW. The control block diagrams for DG1 and DG2 are shown in Figure 19 and Figure 20 respectively. The load changes are similar to that in Table 4.

Figure 21 shows that the power outputs settle to steady-state in about 100–200 s. In comparison, the settling time for DGs in droop control is about 20–40 s. The slower response time in Master-Slave mode is due to the slow integral control of DG1 to bring the system voltage and frequency back to their normal setpoints. The frequency is brought back close to 50 Hz after each disturbance is shown in Figure 22. Similarly, the voltage on bus A is maintained closed to 415 V after each disturbance (as shown in Figure 23). As in previous case, due to the voltage droop characteristics, DG2 consumes reactive power when the voltage is high. This causes DG1 to supply reactive power and hence increases the amount of losses in the line due to a higher current flow through the line. These experiments demonstrate the functionality of EPGC in facilitating testing of grid-related equipment with realistic power magnitudes. It also allows experimental testing of different control strategies and investigation into practical problems that may not be apparent in computer simulated environment.

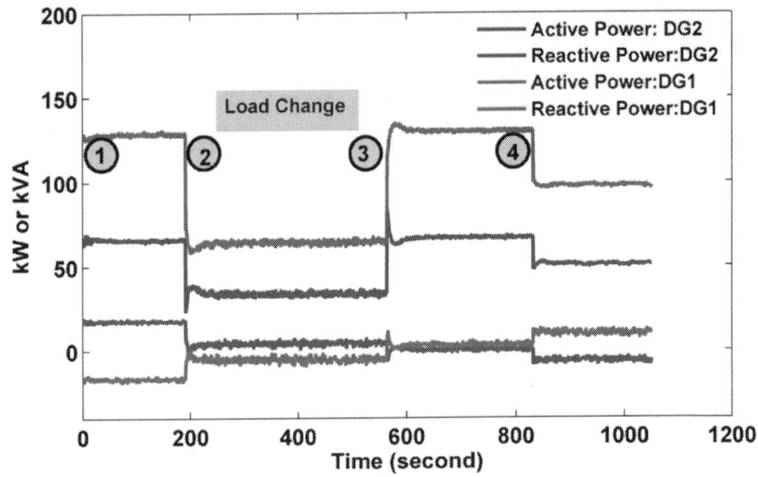

Figure 16. Active and Reactive Power Output of DGs in droop control.

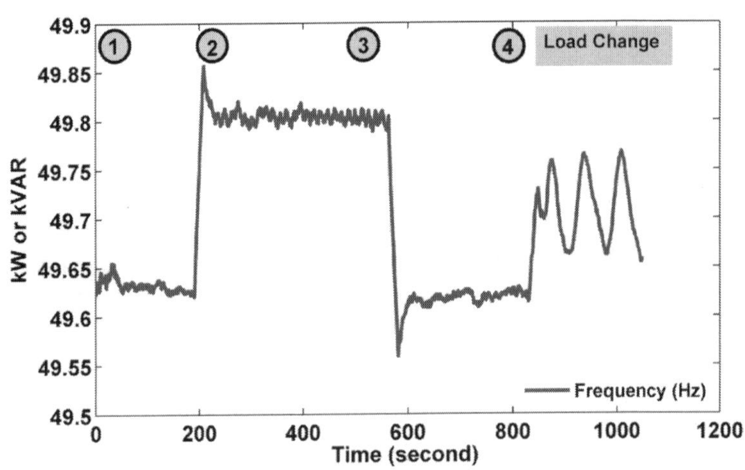

Figure 17. Frequency of DGs in droop mode.

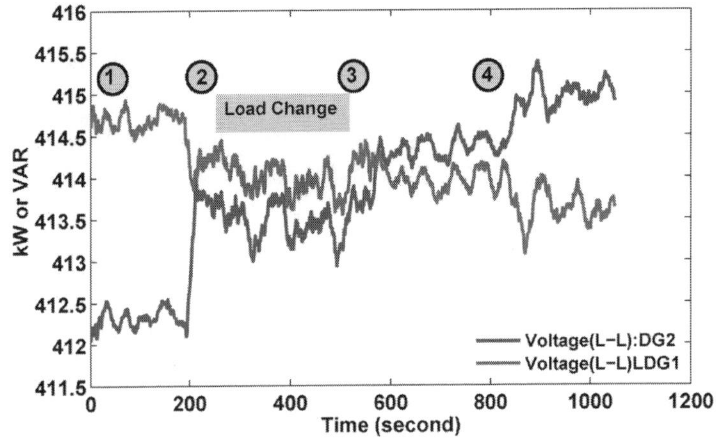

Figure 18. Voltages of DGs in droop mode.

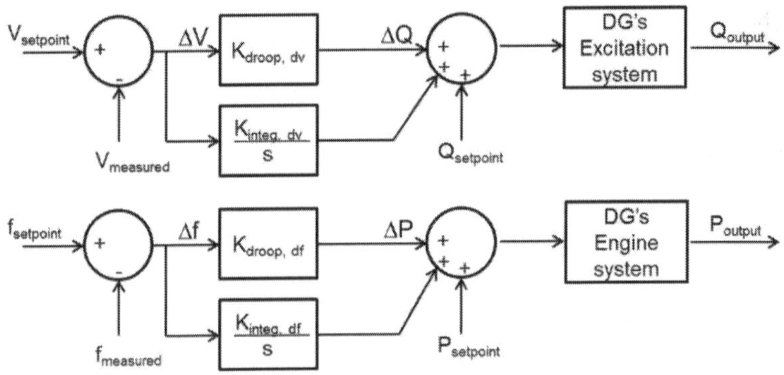

Figure 19. Control block diagrams for DG1.

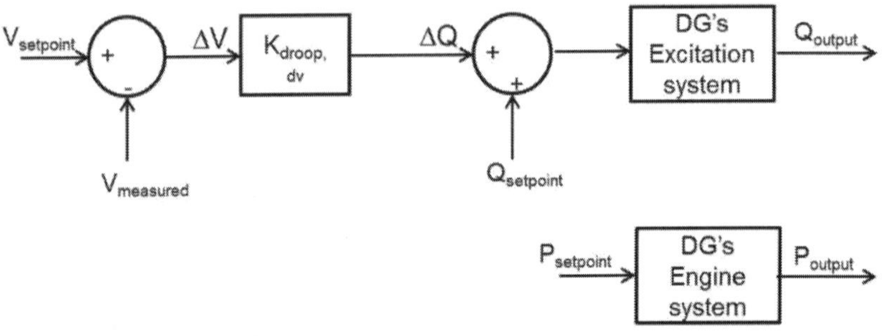

Figure 20. Control block diagrams for DG2.

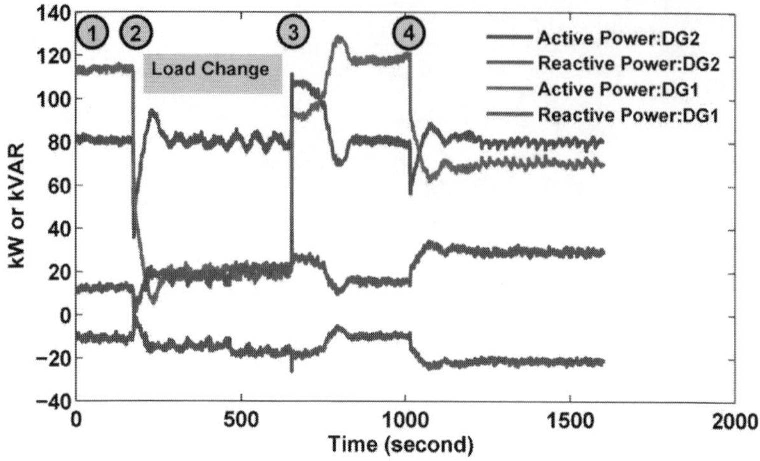

Figure 21. Active and Reactive Power Output of DGs in Master-Slave Control.

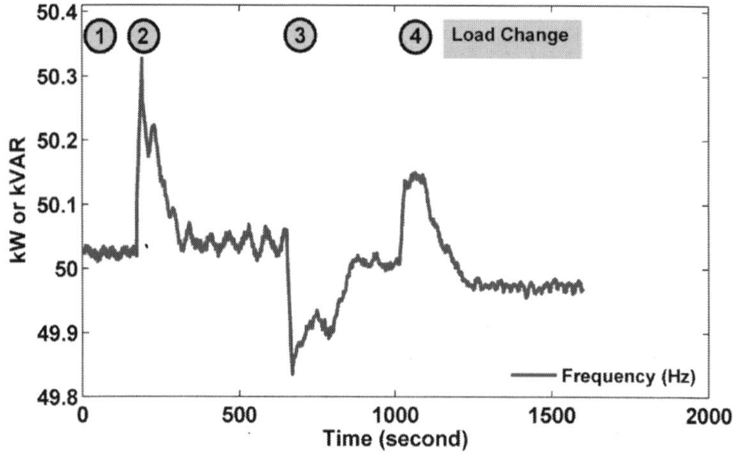

Figure 22. Frequency of DGs in Master-Slave Control.

3.5. Energy Efficiency of Integrated Thermal Grid in Microgrid

An electrical microgrid integrated with thermal grid are noticeable to achieve high efficiency in the range of 60%–85% at design condition depending on the type of thermal supply *i.e.*, steam, hot water or chilled water [27]. During actual operation, diesel generators follow the electrical load, and any variation in the electrical load will influence the diesel generator operation, which in turn affects the exhaust gas properties and thermal grid performance. A detailed experiment and test is required to investigate the performance of the thermal grid at different part load conditions, which arise in the waste heat generation and thermal loads. This case study investigates the energy efficiency of a thermal grid at different part loads faced by thermal systems. Figure 24 shows the heat recovered from the exhaust gas of a 125 kVA generator using a waste heat recovery system at constant hot water flow. The heat energy recovered is 75% to 85% at full load due to high exhaust flow and temperature. At part loads, the heat energy recovered declines all the way to 30% to 35%.

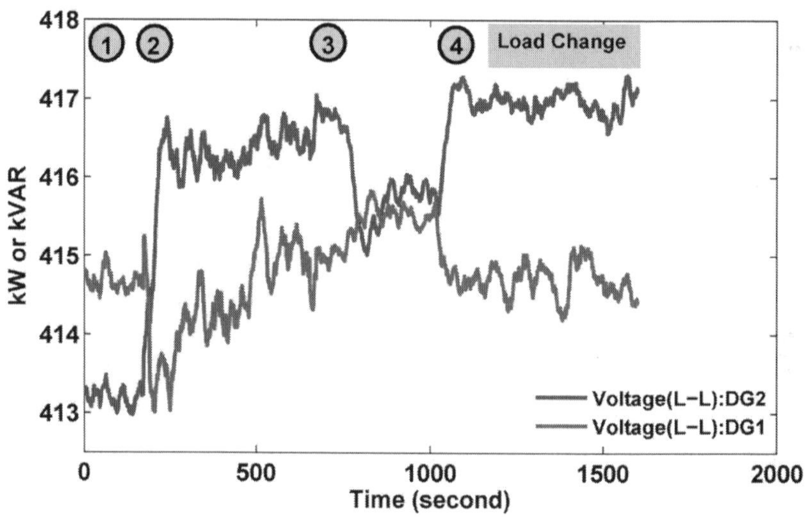

Figure 23. Voltages of DGs in Master-Slave Control.

The temperature of hot water generated by WHRS is at around 70 to 90 ∘C which is more suitable for cleaning and cooking purposes. Hot water can also be used to generate chilled water to support air conditioning purpose. By exploiting thermal storage, the excess hot water is stored and used to peak shaving, shifting operation and during unavailability of exhaust gas conditions. Figure 25 shows the heat availability in the thermal storage, the heat input to the absorption chiller and the cooling capacity generated by the chiller. The heat availability in thermal storage decreases with time because of supplying the hot water to the absorption chiller. The area of heat input and cooling capacity curves shows 30%–50% of heat input is converted to cooling capacity in the chilled water to support air conditioning loads. The range 30%–50% is in agreement for the single-effect hot water driven chiller at part load conditions [29]. This four cubic meter thermal storage can support chiller operation for more than an hour without WHRS operation. However, the cooling capacity generation is reduced due to a decrease in the heat capacity and heat input.

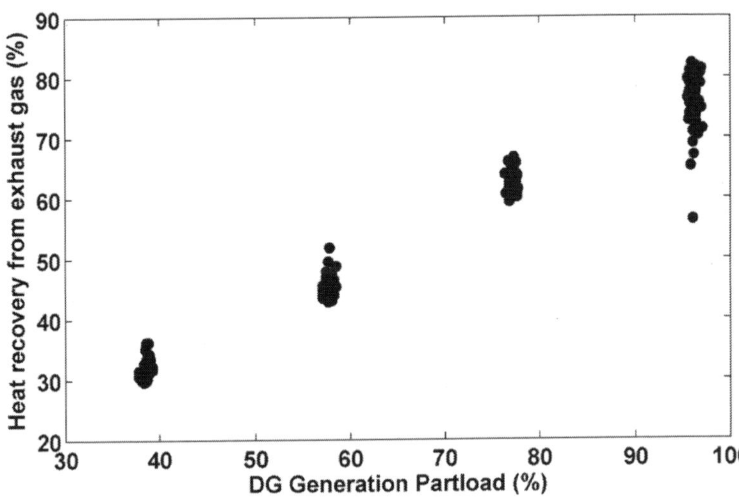

Figure 24. Heat recovered from the exhaust gas at different part loads of diesel generator.

The performance of the absorption chiller is influenced by the hot water, cooling water and chilled water inlet temperature and flow [30]. At rated condition, the hot water, cooling water and chilled water requirement is 90, 29.5 and 12 °C. Any variation in the inlet temperature or flow due to generator operation, thermal loads and weather will affect the chiller performance. Figure 26 shows the performance of the chiller with the variation in the hot water inlet temperature at constant flow. The decrease in the hot water temperature from the design condition affects the efficiency of the chiller. Similarly chiller efficiency deteriorates when chilled water inlet temperature decreases or cooling water inlet temperature increases from the design value.

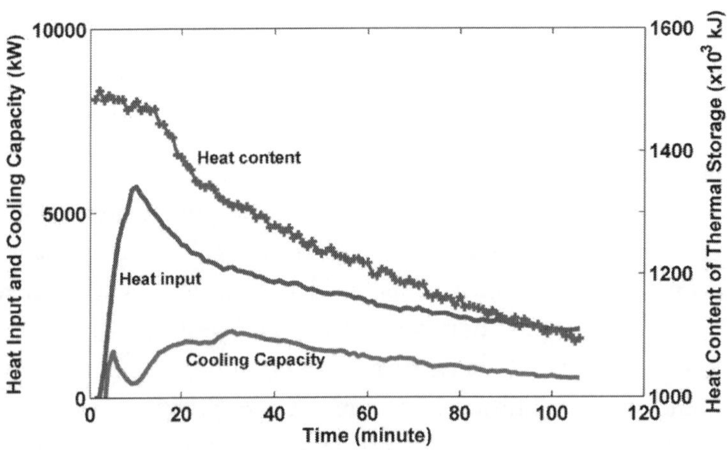

Figure 25. Cooling capacity generation and hot water temperature in the thermal storage.

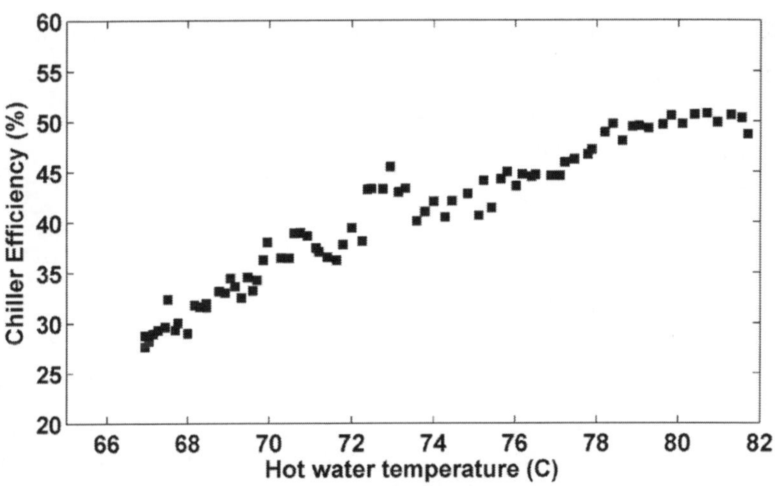

Figure 26. Chiller Efficiency as Coefficient of performance.

Figure 24, Figure 25 and Figure 26 confirm that the performance of thermal systems is significantly influenced by the generators, thermal storage, chiller operation and thermal loads. The operational decisions are the key to achieve optimized performance and suppress performance deterioration. An advanced real-time optimized control is developed and implemented to derive optimized operational decisions every 10 to 30 min based on the thermal grid variables, predicted thermal load and weather condition. This advanced control is a remote system connected to the thermal grid energy management system to read the thermal grid variables and write the control decisions respectively.

The developed control system takes the (i) exhaust gas properties, (ii) hot water, chilled water and cooling water temperature and (iii) weather properties as inputs from the thermal grid energy management system and optimizes the thermal grid performance with respect to kWRT (as shown in Equation (5)). The optimized operational decisions derived are flowrate of hot water, chilled water and cooling water, and the air flow in cooling tower. These decisions are implemented in the testing facility by revising the corresponding setpoints of variable speed drives of hot water pump, chilled water pump, cooling water pump and cooling tower fan. The thermal grid performance is described in (Equation (5)) as kWRT which is the ratio of power consumption to generate one refrigeration ton of cooling capacity. The power consumption refers to the sum of power consumed by the chiller (PCH), cooling tower (PCT) and the pumps in hot, chilled and cooling water circuits (Phw, Pchw, Pcw). The verified thermodynamic model of thermal grid is used in the optimized control to generate the operational decisions.

$$Min : kWRT = \frac{Power\ consumed\ (kW)}{cooling\ capacity\ generated\ (RT)}$$
$$kWRT = \frac{\sum_{i=hw,chw,cw,CH,CT} P_i}{m_{chw} C_p \Delta T} \tag{5}$$

Figure 27 shows the energy efficiency of thermal grid, a high efficiency system achieves a smaller value of kW per ton to generate cooling capacity. Figure 27 shows the preliminary results of thermal grid performance which show that optimized control scheme have performance benefits of up to 20% at part load conditions. The performance benefits are projected up to 40% in large capacity system in the range of 500 KVA diesel generators [28]. This thermal grid facility with adequate instrumentation allows system level analysis and validate new optimized control algorithms.

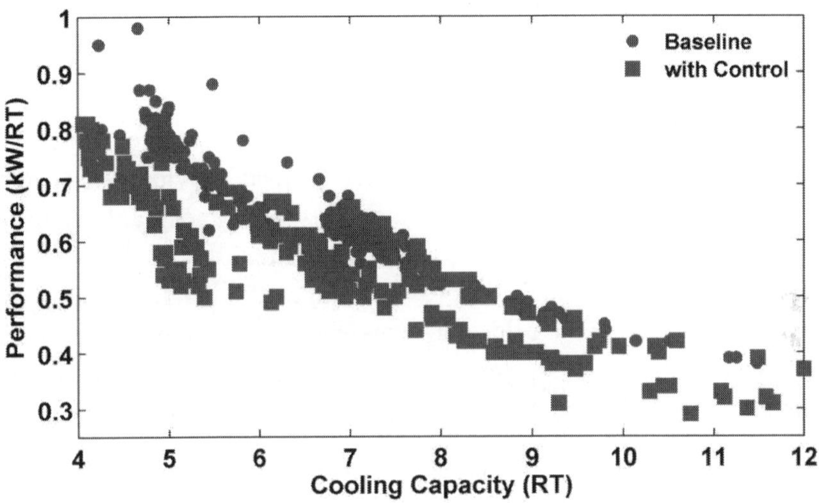

Figure 27. Thermal grid performance with baseline operation and optimized controller.

4. CONCLUSIONS

This paper presented an overview of EPGC's microgrid facility which has flexible grid architecture and variety of emulators and distributed energy sources. The facility allows testing and development of microgrid control strategies in different grid configurations and modes of operation. In addition, the facility also has a thermal grid, which allows for experimentation of integrated electrical-thermal systems to optimise and achieve higher energy efficiency and cost benefits. The functionalities of the microgrid test facility have been highlighted through selected case studies. The testing of DG and future microgrid technologies with realistic power magnitudes and accordance with the international standards make this test facility a unique and key enabler for the widespread adoption of the microgrid. The experimental verification provides higher confidence in the technology which is surely limited in the computer simulation environment. The influence of physical limits in the electrical and thermal systems and their response time can be precisely observed in the experiments, which is not evident in software models and simulators.

AUTHOR CONTRIBUTIONS

Sundar Raj Thangavelu: Developed the optimized control algorithm for thermal grid and validated the performance provided in Section 3.5. Inam Ullah Nutkani: Contributed the experiment and analysis presented in Section 3.1 to Section 3.3. Chia Meng Hwee: Contributed the experiments and analysis presented in Section 3.4. Aung Myat: Designed the thermal grid and conducted experiments discussed in Section 3.5. Ashwin Khambadkone: Scoped the experiments and framed this article structure. All authors contributed in scoping and drafting relevant sections of this article.

CONFLICTS OF INTEREST

The authors declare no conflict of interest.

REFERENCES

1 *Microgrids-Benefits, Models, Barriers and Suggested Policy Initiatives for the Commonwealth of Massachusetts*; KEMA Inc.: Burlington, MA, USA, 2014.

2 Ackermann, T.; Andersson, G.; Soder, L. Distributed generation: A definition. *Electr. Power Syst. Res.* **2001**, *57*, 195–204.

3 Hossain, E.; Kabalci, E.; Bayindir, R.; Perez, R. Microgrid testbeds around the world: State of art. *Energy Convers. Manag.* **2014**, *86*, 132–153.

4 Ramazan, B.; Eklas, H.; Erdal, B.E.K. Microgrid Facility at European Union. In Proceedings of the 3rd International Conference on Renewable Energy Research and Applications, Milwakuee, WI, USA, 19–22 October 2014.

5 Morozumi, S. Micro-grid Demonstration Projects in Japan. In Proceedings of the Power Conversion Conference, Nagoya, Japan, 2–5 April 2007; pp. 635–642.

6 *International Microgrid Assessment: Governance, INcentives, and Experience (IMAGINE)*; European Council for an Energy-Efficient Economy: Hyeres, France, 2013.

7 Microgrid Deployment Tracker. Available online: https:// www.navigantresearch.com/research/microgrid-deployment-tracker-2q15 (accessed on 8 September 2015).

8 Christopher, V.D.E. Microgrids: A Regulatory Perspective. Available online: http://www.cpuc.ca.gov/ NR/rdonlyres/ 01ECA2 96-5E7F-4C23-8570-1EFF2DC0F278/0/PPDMicrogridPaper 414. pdf(accessed on 7 September 2015).

9 IEEE Standard 1547-2003. In *IEEE Standard for Interconnecting Distributed Resources with Electric Power Systems*; IEEE: Piscataway, NJ, USA, 2003; pp. 1–28.

10 IEEE Standard 1547.4-2011. In *IEEE Guide for Design, Operation, and Integration of Distributed Resource Island Systems with Electric Power Systems*; IEEE: Piscataway, NJ, USA, 2011.

11 Report on standards and grid code requirements applied to LV DG devices. Available online: http://www.microgrids.eu/ documents/ 654.pdf (accessed on 7 September 2015).

12 Boemer, J.; Burges, K.; Kumm, T. Compliance with Technical Codes Turns into Precondition for Support and System Services Bonus for Wind Power Plants in Germany. In Proceedings of the

2009 IEEE Bucharest PowerTech, Bucharest, Rumania; 2009; pp. 1–8.

13 Brooks, D.; Patel, M. Panel: Standards Amp, Interconnection Requirements for Wind and Solar Generation NERC Integrating Variable Generation Task Force. In Proceedings of the 2010 IEEE PES Transmission and Distribution Conference and Exposition, New Orleans, LA, USA, 19–22 April 2010.

14 *Design and operational Recommendations on Grid Connection of PV Hybrid Mini-Grids*; International Energy Agency: Paris, France, 2011.

15 Standard Code: JEAC 9701-2012. In *Grid Interconnection Code*; Japan Electric Association: Tokyo, Japan, 2012.

16 *Technical Requirements Guideline of Grid interconnection to Secure Electricity Quality*; Ministry of Economy, Trade and Industry (METI): Tokyo, Japan, 2013.

17 Kish, G. Addressing Future Grid Requirements for Distributed Energy Resources. Master's Thesis, University of Toronto, Toronto, ON, Canada, 2011.

18 Kish, G.; Lehn, P. Microgrid Design Considerations for Next Generation Grid Codes. In Proceedings of the 2012 IEEE Power and Energy Society General Meeting, San Diego, CA, USA, 22–26 July 2012; pp. 1–8.

19 El-Hawary, M.E. The Smart Grid-State-of-the-Art and Future Trends. *Electr. Power Compon. Syst.* **2014**, *42*, 239–250.

20 Sharma, G.; Xie, L.; Kumar, P.R. On the Optimality of De-Synchronized Demand Response with Stochastic Renewables and Inertial Thermal Loads. In Proceedings of the 2013 IEEE 52nd Annual Conference on Decision and Control, Firenze, Italy, 10–13 December 2013; pp. 6292–6297.

21 Sanaye, S.; Sarrafi, A. Optimization of combined cooling, heating and power generation by a solar system. *Renew. Eenergy* **2015**, *80*, 699–712.

22 Diesel Generators. Available online: https://www.fgwilson.com/ (accessed on 7 September 2015).

23 High Performance Solar Modules, REC Peak Energy Series. Available online: http://www.recgroup.com (accessed on 7 September 2015).

24 Murphy, M.D.; O'Mahony, M.J.; Upton, J. Comparison of control systems for the optimisation of ice storage in a dynamic real time electricity pricing environment. *Appl. Energy* **2015**, *149*, 392–403.

25 Bari, S.; Hossain, S.N. Waste heat recovery from a diesel engine using shell and tube heat exchanger. *Appl. Therm. Eng.* **2013**, *61*, 355–363.

26 *CRC Handbook of Energy Efficiency*; Frank Krieth, R.E.W., Ed.; CRC Press: Boca Raton, FL, USA, 1996.

27 Mohan, G.; Dahal, S.; Kumar, U.; Martin, A.; Kayal, H. Development of natural gas fired combined cycle plant for tri-generation of power, cooling and clean water using waste heat recovery: techno-economic analysis. *Energies* **2014**, *7*, 6358–6381.

28 Thermax Multi Energy Absorption Chillers for Small Scale Cogeneration Projects. Available online: http://www.districtenergy.org./pdfs/2012CHP-Texas (accessed on 7 September 2015).

29 Thermax products, Absorption Systems, Absorption Chillers. Available online: http://www.thermaxglobal.com(accessed on 7 September 2015).

30 Prasartkaew, B. Performance Test of a Small Size LiBr-H_2O Absorption Chiller. *Energy Procedia* **2014**, *56*, 487–497.

CHAPTER 9

Optimized Adaptive Perturb and Observe Maximum Power Point Tracking Control for Photovoltaic Generation

Luigi Piegari [1,†], Renato Rizzo [2,†], Ivan Spina [2,*] and Pietro Tricoli [3,†]

[1] Department of Electronics, Information and Bioengineering, Politecnico di Milano, Piazza Leonardo da Vinci 32, 20133 Milan, Italy

[2] Department of Electrical Engineering and Information Technologies, University of Naples Federico II, Via Claudio 21, 80125 Naples, Italy

[3] School of Electronic, Electrical and Computer Engineering, University of Birmingham, Gisbert Kapp Building, Birmingham B15 2TT, UK

ABSTRACT

The power extracted from PV arrays is usually maximized using maximum power point tracking algorithms. One of the most widely used techniques is the perturb & observe algorithm, which periodically perturbs the operating point of the PV array, sometime with an adaptive perturbation step, and compares the PV power before and after the perturbation. This paper analyses the most suitable perturbation step to optimize maximum power point tracking performance and suggests a design criterion to select the parameters of the controller. Using this proposed adaptive step, the MPPT perturb & observe algorithm achieves an excellent dynamic response by adapting the perturbation step to the actual operating conditions of the PV array. The proposed algorithm has been validated and tested in a laboratory using a dual input inductor push-pull converter. This particular converter topology is an efficient interface to boost the low voltage of PV arrays and effectively control the power flow when input or output voltages are variable. The experimental results have proved the superiority of the proposed algorithm in comparison of traditional perturb & observe and incremental conductance techniques.

KEYWORDS

Maximum Power Point Tracking (MPPT); Perturb & Observe (P&O); photovoltaic generation; push-pull converter

1. INTRODUCTION

Photovoltaic (PV) power generation has received significant attention from scientists over the last few years to help reduce the environmental pollution inherently associated with traditional electric generators. The economic convenience of PV generation is directly connected to the cost of the cells and the amount of energy that the arrays are capable of supplying over their life. The former is influenced mainly by the price of the feedstock for the PV industry and the improvement of fabrication technologies. The latter is connected to the efficiency of the power conversion system, which is normally necessary when PV arrays are used for power generation. At the present state of the art, the control algorithms used for the maximization of the power extracted from PV arrays are widely known as maximum power point tracking (MPPT) algorithms. These power conditioners guarantee that PV arrays operate close to their point of maximum efficiency under any weather conditions. This is particularly important for variable irradiation levels, because the efficiency of the PV modules is very low when the operating point is far from the maximum power point.

The MPPT techniques proposed so far in the technical literature can be classified mainly as Perturb & Observe (P&O), Incremental Conductance (IC), and Temperature Gradient (TG) techniques [1]. The P&O algorithm is very popular because of its easy implementation [2,3,4,5]. The operating voltage of the converter is changed slightly at every control step and the consequent variation of the power output is measured. If the perturbation has caused a power increase, the voltage is further changed in the same direction; if not, a backwards perturbation is applied. A typical disadvantage of this technique is that the operating point always oscillates around the maximum, losing some part of the energy actually available [6]. Smaller perturbation steps can mitigate the problem, although the algorithm becomes slow at finding the new maximum when irradiation conditions change suddenly.

Several improvements on the P&O algorithm have been proposed in order to reduce the oscillations around the maximum power point (MPP) at steady-state, but all of them slow down the response speed of the algorithm when atmospheric conditions change and reduce the efficiency during cloudy days [7]. An adaptive MPPT algorithm for a field-programmable gate array has been proposed in [8]. This adjustable algorithm is based on an improved P&O method that does not need external sensory units. Despite its cost-effectiveness and high-speed processing, the system performance has been demonstrated to be outstanding.

In the original MPPT P&O controller, the adjustment of the operating point is achieved by changing the reference voltage of the controller. However, this adjustment can be made through the duty ratio of the converter such as described in [8]. The experimental results reported in the recent technical literature show that the adaptive algorithm has better performance than conventional P&O techniques in terms of dynamics and efficiency. In particular, an efficiency of the algorithm above 96% has been reported in [9,10]. In [10] several adaptive perturbation functions are considered and the final choice is made by randomly setting the coefficients for each function and comparing the results obtained from numerous tests. Since the function associated with best results shows stability problems in the right side of the maximum power point, the algorithm has been modified by associating zero values for the unstable operation range. In [11] the authors investigated and compared the characteristics and performance of several P&O techniques, showing that some are not truly adaptive, though they present improved performance in comparison to their fixed perturbation step counterparts [12,13,14]. Other techniques propose truly adaptive modifications, but suffer from a high computational load caused by aggressive calculation of derivatives [15,16] and need initial user-dependent constants for the perturb adaptation [17,18]. Novel techniques show improved performance using nonlinear equations [19], fuzzy logic [20], and complicated optimization algorithms [21,22,23].

This paper proposes a design criterion to choose the adaptive perturbation step, in order to assure the most suitable perturbation amplitude for a P&O MPPT algorithm in different operating conditions, while keeping a good dynamic response. In addition to the work developed in [10], this paper proposes a modified P&O algorithm with a properly chosen perturbation function. Experimental tests on the proposed algorithm are carried out on a dual input inductor push-pull converter (DIIPPC) [24] with the aim of comparing the results with both classic P&O and incremental conductance (IC) algorithms. DIIPPCs achieve a high boost ratio and provide isolation between the PV arrays and the load, optimising the power transfer for every irradiation and temperature level. Moreover, these converters present lower output voltage ripple, lower voltage across the main switches, and a lower volt-ampere rating of the transformer than current-fed push-pull converter [24].

This paper is organised as follows: Section 2 presents the DIIPPC used for PV generation and its basic operating characteristics; Section 3 describes the adaptive P&O MPPT algorithm; Section 4 reports the parameter identification of PV panel used for the experiments; Section 5 provide details about the MPPT implementation; Section 6 discusses the experimental validation of the proposed technique and the testing methodologies; Section 7 introduces a discussion on the MPP estimation error.

2. DUAL INDUCTOR PUSH-PULL CONVERTER

The basic configuration of a DIIPPC [24] is illustrated in Figure 1. DIIPPCs operate similarly to converters with a single inductor.

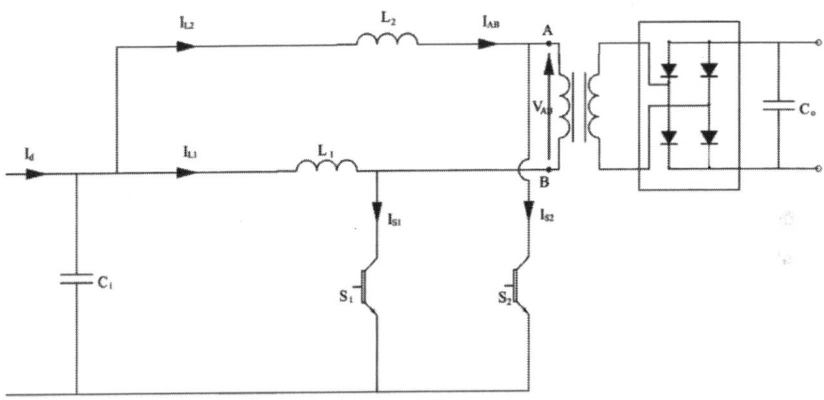

Figure 1. Circuit diagram of the dual input inductor push-pull converter.

The switches of the input stage are both turned on to charge the input inductors and are alternately turned off to transfer the energy to the primary winding of transformer. The two components are driven with the same control signal delayed by one half of the period. Therefore, a duty cycle greater than 0.5 is sufficient to guarantee the operating principle just described.

Figure 2 shows the control signals for the switches S1 and S2, the current and the voltage across the transformer, the source current, input inductance currents and the current across one of the main switches for a typical operating condition. In order to make the converter suitable to be used for interfacing PV panels, according with [25,26] an input capacitor is connected. In the hypothesis that this capacitance is high enough, during one switching period the input voltage can be considered constant.

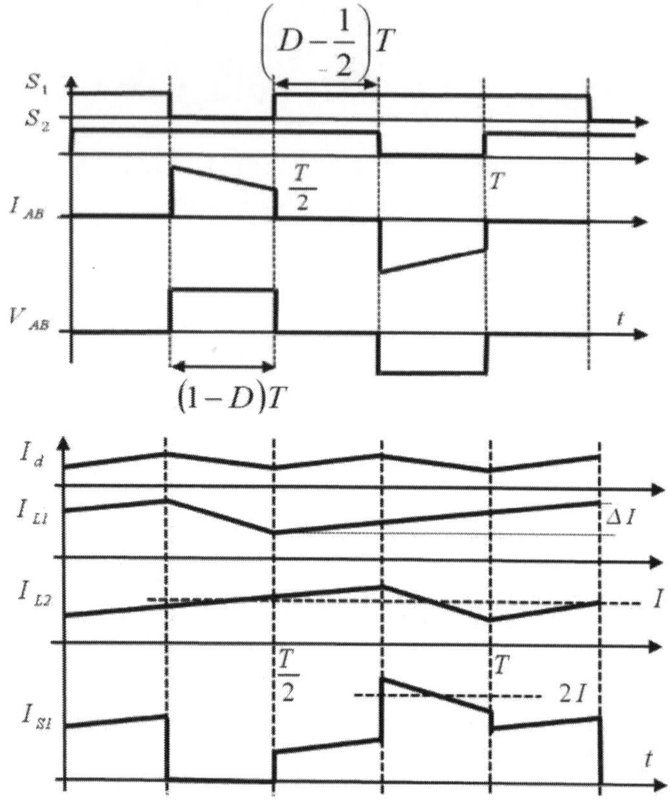

Figure 2. Control signals and primary current and voltage across the transformer (upper diagram); source current, inductor currents and current across the first switch (lower diagram).

The output voltage of this converter is increasing with the duty cycle of the switches, because the input stage acts as a boost converter. The output voltage of the converter can be expressed as:

$$V_0 = \frac{N_2}{N_1} \frac{V_i}{1-D} \tag{1}$$

where V_i is the input voltage, D is the duty cycle (defined as the ratio between the on-state time and the switching period, T_{on}/T), and N_1 and N_2 are the numbers of turns of the primary and secondary windings, respectively.

The rapid change of the inductor current, due to the turn-off of the switch, causes an overvoltage across the switch itself. For this reason, a snubber circuit may be added to reduce the overvoltage. Since the efficiency of the converter is not the main focus of the paper, a traditional dissipative snubber has been used.

3. ADAPTIVE PERTURBATION FUNCTION FOR P&O ALGORITHM

The MPPT algorithm proposed by the authors in [10] is a variation of the classic P&O method and has been set up and implemented for the control of the DIIPPC. The main limitation of the P&O algorithm is the compromise between a fast dynamic response and a reduced oscillation at steady-state. Large perturbation steps imply a fast dynamic performance, whereas small perturbation steps ensure little oscillations at steady-state. In particular conditions, during cloudy days, the solar irradiation can vary many times and quickly. This situation can cause fault or reduction of efficiency in MPP trackers. In order to address this problem some studies has been proposed in the technical literature [27,28]. In these conditions, it is necessary to achieve a good dynamic response to avoid losing a great quantity of energy. The efficiency of the P&O technique depends on the perturbation step. It is particularly difficult to find the optimal perturbation step, because this depends on the panel characteristics, working conditions and the converter's topology. To obtain a faster dynamic with small oscillations without knowing accurately the technical details of the PV system, the proposed adaptive P&O method operates with variable perturbation step adapted to the actual operating conditions. In particular, when the operating point is far from the MPP, large perturbation steps are chosen, whereas small ones are used in proximity to the maximum. Therefore, the algorithm needs to know the MPP position. However, a precise knowledge is not required because the P&O technique always works around the real maximum.

In the proposed method, the perturbation step is defined by the following equation:

$$\Delta V(V) = f\left(V_{\mathrm{M,est}} - V\right) + \Delta V_{\mathrm{min}} \tag{2}$$

where ΔV is the perturbation step; $V_{\mathrm{M,est}}$ is the estimated voltage at MPP; V is the measured voltage; ΔV_{min} is the minimum value of the perturbation.

In order to guarantee a high dynamic performance and limited steady-state oscillations, the function f must be a positive function and must assume a high value when the working point is far from the maximum; moreover, it must be nearly zero in proximity to the maximum. The condition of working in proximity to the MPP is assured by setting ΔV_{min} at an appropriate value. If the working point is near to the value of $V_{\mathrm{M,est}}$, $\Delta V \approx \Delta V_{\mathrm{min}}$; when $V \equiv V_{\mathrm{M,est}}$, $f(0) = 0 \rightarrow \Delta V(0) = \Delta V_{\mathrm{min}}$.

In the previous work [10], $f(V_{\mathrm{M,est}} - V)$ has been chosen to be a quadratic function in the left side of the MPP, and zero in the right side of MPP for stability problems. This procedure for choosing the function given by Equation (2) has not been fully justified and no correlation between $f(V_{\mathrm{M,est}} - V)$ and ΔV_{min} has been considered. In this paper a criterion to properly choose the adaptive perturbation function is discussed taking into account the possible correlation between the two addends of Equation (2) and the estimation error in evaluating the MPP.

If the MPP voltage estimation were perfect ($V_{\mathrm{M,est}} \equiv V_{\mathrm{M}}$, where V_{M} is the actual position of the MPP), and the converter did not introduce delays, the best expression for f and ΔV_{min} would be:

$$\begin{aligned} f\left(V_{\mathrm{M,est}} - V\right) &= \left|V_{\mathrm{M,est}} - V\right| \\ \Delta V_{\mathrm{min}} &= 0 \end{aligned} \tag{3}$$

In this ideal condition, the MPP would be reached after one iteration. However, the MPP voltage estimation introduces an error. Therefore, in order to make the algorithm capable of reaching the actual MPP, it is necessary to have $\Delta V_{min} \neq 0$. In addition, when the measured voltage V is close to V_M, the function f should yield a value lower than ΔV_{min} when $V \equiv V_M$:

$$f\left(V_{M,est} - V_M\right) < \Delta V_{min} \tag{4}$$

The condition given by Equation (4) guarantees that f helps to increase the rapidity of the MPPT algorithm far from the maximum power point, but does not worsen the oscillations when the PV array is operating around V_M. Indicating with $k$the maximum estimation error in p.u. it is:

$$\frac{\left|V_{M,est} - V_M\right|}{V_M} \leq k \tag{5}$$

The algorithm has to be capable of correctly matching the MPP also in case of wrong estimation of V_M. For this reason, a minimum perturbation step has to be introduced, as reported in Equation (2). Being k the maximum allowed estimation error, the algorithm could go to work in a point far from the real MPP of, at worst, kV_M. In order to achieve the correct matching of the MPP it has to be:

$$\Delta V_{min} > kV_M \tag{6}$$

In this way, the P&O action can compensate the estimation error in two steps.

Equation (6) proves that the linear expression (3) cannot be used in practice because estimation errors would be too high to ensure small oscillations at steady state. For this reason, polynomial expressions with higher order or exponential expressions are preferred in the practical implementation of the algorithm. Choosing, for example, a quadratic function:

$$f\left(V_{\mathrm{M,est}}-V\right)=a\left(V_{\mathrm{M,est}}-V\right)^{2} \tag{7}$$

It is possible to set a high value for f in order to have a quick answer when the voltage is far from the MPP. In particular, when V is zero (the furthest position), the MPP can be reached in two steps if the function f is selected as:

$$f\left(V_{\mathrm{M,est}}\right)=0.5V_{\mathrm{M,est}} \tag{8}$$

This last equation, together with Equation (7), gives the value of parameter a:

$$a=\frac{0.5}{V_{\mathrm{M,est}}} \tag{9}$$

Substituting the expression of f (given by Equation (7)) into Equation (4), and using Equation (9) for parameter a, it is possible to obtain the following condition on ΔV_{min}:

$$\Delta V_{\mathrm{min}}>\frac{0.5}{V_{\mathrm{M,est}}}\left(V_{\mathrm{M,est}}-V_{\mathrm{M}}\right)^{2}=A \tag{10}$$

where the quantity A systemizes the right term of the inequality. Moreover, from Equation (5) the following condition can be obtained:

$$A = \frac{0.5}{V_{\mathrm{M,est}}}\left(V_{\mathrm{M,est}} - V_{\mathrm{M}}\right)^2 \leq \frac{0.5}{V_{\mathrm{M,est}}} V_{\mathrm{M}}{}^2 k^2 = B \tag{11}$$

where the quantity B synthesizes the right term of the inequality. In order to satisfy condition (10) it is possible to choose $\Delta V_{\min}$ higher than B, *i.e.*, if $\Delta V_{\min} > B \Rightarrow \Delta V_{\min} > A$ (since $A \leq B$):

$$\Delta V_{\min} > B \quad \Rightarrow \quad \begin{cases} \Delta V_{\min} > A \\ \text{with:} A \leq B \end{cases} \tag{12}$$

Thus, $\Delta V_{\min} > B$ is a sufficient (but not necessary) condition to satisfy Equation (10). Solving Equation (5) it is possible to deduce the following conditions on $V_{\mathrm{M,est}}$, and, consequently on the quantity B:

$$\frac{\left|V_{\mathrm{M,est}} - V_{\mathrm{M}}\right|}{V_{\mathrm{M}}} \leq k \Rightarrow \begin{cases} V_{\mathrm{M,est}} \in \left]V_{\mathrm{M}}, (1+k)V_{\mathrm{M}}\right] \\ \text{or} \\ V_{\mathrm{M,est}} \in \left[(1-k)V_{\mathrm{M}}, V_{\mathrm{M}}\right[\end{cases} \Rightarrow V_{\mathrm{M,est}} \geq (1-k)V_{\mathrm{M}} \Leftrightarrow B \leq \frac{0.5k^2}{(1-k)} V_{\mathrm{M}} = C \tag{13}$$

where the quantity C synthesizes the right term of the last inequality. In order to satisfy $\Delta V_{\min} > B$ it is possible to choose $\Delta V_{\min}$ higher than C, *i.e.*, if $\Delta V_{\min} > C \Rightarrow \Delta V_{\min} > B$ (since $B \leq C$):

$$\Delta V_{\min} > C \quad \Rightarrow \quad \begin{cases} \Delta V_{\min} > B \\ \text{with:} B \leq C \end{cases} \tag{14}$$

which, substituting the expression of the term *C*, gives:

$$\Delta V_{\min} > 0.5 \frac{k^2}{(1-k)} V_{\mathrm{M}} \tag{15}$$

Equation (15) returns a minimum variable step of 0.56% for a 10% estimation error, which ensures reasonably small oscillations at steady state.

4. PV PANEL IDENTIFICATION

The PV array is composed of 36 cells of monocrystalline silicon wafers, connected in series and measuring a total active area of 0.54 m^2. The technical data of the PV array are reported in Table 1. The PV array model is mathematically described by the following equations [10]:

$$\begin{aligned} V &= \frac{AKT}{q} \ln\left(\frac{I - I_{\mathrm{ph}} - I_{\mathrm{sat}}}{I_{\mathrm{sat}}}\right) - r_s I \\ I_{\mathrm{ph}} &= \left[I_{\mathrm{scr}} + k_i \left(T - T_{\mathrm{r}}\right)\right] s \\ I_{\mathrm{sat}} &= I_{\mathrm{rr}} \left(\frac{T}{T_{\mathrm{r}}}\right)^3 \exp\left[\frac{q E_{\mathrm{g}}}{K}\left(\frac{1}{T_{\mathrm{r}}} - \frac{1}{T}\right)\right] \end{aligned} \tag{16}$$

where I_{scr} is the short circuit current, I_{rr} is the reverse current at the reference temperature, *q* is the electronic charge, *A* is the junction ideality factor, *K* is the Boltzmann constant, *T* is the absolute temperature of the array, T_r is a constant equal to 273.14 K, r_s is the series resistance, s is the p.u. irradiation, E_g is the band gap energy of the semiconductor, and k_i is the short-circuit current temperature coefficient.

The curves of the PV array have been evaluated experimentally for several temperatures, ranging from 23 °C to 40 °C with a step of 1 °C. However, only some of these curves are presented in Figure 3 for a better visualisation. The tests were made with an artificial light to control the level of irradiance and keep the irradiance smooth during the test. The open circuit voltage of the panel, V_{oc}, presents a decreasing rate of 2.3 mV/°C, whereas the short-circuit current, I_{scr}, presents an increasing rate of 0.2%/°C. This is reflected by a power peak reduction, which is equal to 0.6%/°C–0.7%/°C. The experimental data have been fitted using a least mean square numeric procedure in order to obtain the parameters of the model. The results of this procedure are shown in Table 2. The curves obtained by the model are also plotted in Figure 3, showing very good agreement with the experimental data.

Table 1. Characteristics of PV array.

Type	Aluminum-Tedlar125M36
Open circuit voltage, V_{oc}	21.85 V
Short circuit current, I_{sc}	4.52 A
Maximum power, P_M	69.03 W
Maximum power voltage, V_M	17.74 V
Maximum power current, I_M	3.89 A

Table 2. Main data of the PV panel used.

A [–]	I_{rr} [nA]	r_s [Ω]	I_{scr} [A]	k_i [A/K]
57	−15.8	0.25	3.5	0.001

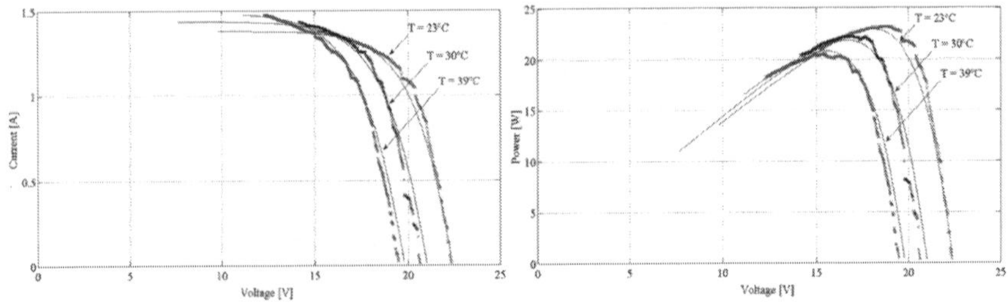

Figure 3. PV current and power curves for different array temperatures and their fitting.

5. ALGORITHM IMPLEMENTATION

The MPPT algorithm presented in this paper requires the estimation of the maximum V_M. Once the panel has been identified according to the procedure of the previous section, the theoretical position of V_M can be determined by the mathematical model of the panel. In practical applications the characteristic of the panel changes during its life for ageing and dust. For this reason, a periodical characterization has to be performed and the experimental characteristic has to be registered. A good time interval for characterization of the panel can be one month. During one month the efficiency degradation of a PV panel has been estimated to be 7% [29]. Anyway, with the proposed algorithm, an error in the estimation of the theoretical maximum is compensated by the P&O function. As analyzed in Section 7 a 10% error in the estimation of V_M does not practically affect the performances of the proposed algorithm. It is worth noting that, the chosen converter, allows the complete catching of panel characteristic because it can make the panel work from short circuit (duty cycle equal to 1) to open circuit (duty cycle equal to 0).

In order to estimate the MPP voltage and current, it is necessary to measure the irradiation and the temperature. Although it is quite easy and inexpensive to measure the temperature, the measurement of the irradiation greatly increases the cost of the MPPT control system. For this reason, the power-voltage characteristic has been preferred to the power-current characteristic. Indeed, the voltage at the MPP is weakly variable with the irradiation, whereas the current at the MPP is weakly variable with the temperature. In the following figures, the irradiation is normalised to 1 kW/m^2. Figure 4shows the MPP voltage as a function of the irradiation for different temperatures.

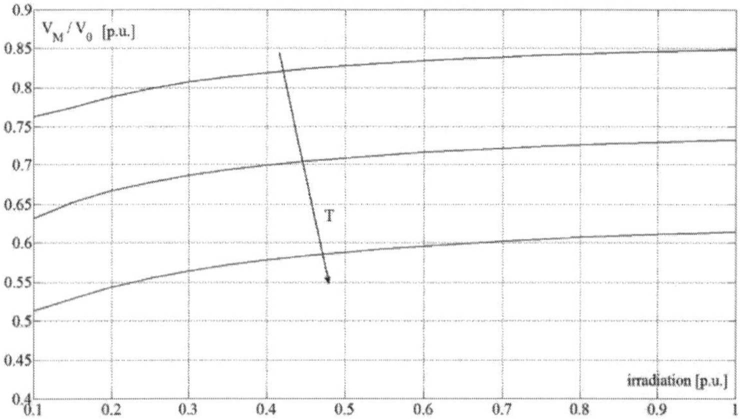

Figure 4. MPP voltage *vs.* irradiation for different temperatures.

The power-voltage characteristic has been chosen for the algorithm implementation, in order to use only one temperature sensor for the individuation of the maximum position. Using the parameters of the panel, reported in Table 2, the error committed in the MPP voltage estimation is reported in Figure 5.

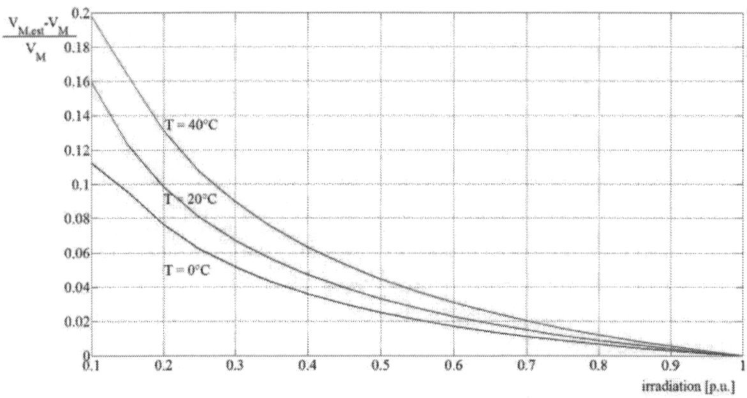

Figure 5. MPP voltage estimation error in p.u. *vs.* irradiation for different temperatures.

Figure 5 shows that when the irradiation is higher than 0.3 p.u. the MPP voltage estimation error is lower than 10%. Without a P&O algorithm, a MPP voltage estimation error implies a power loss with respect to the maximum power that can be generated. Figure 6 shows this power loss in per unit of the nominal array power as a function of the p.u. irradiation for different temperatures.

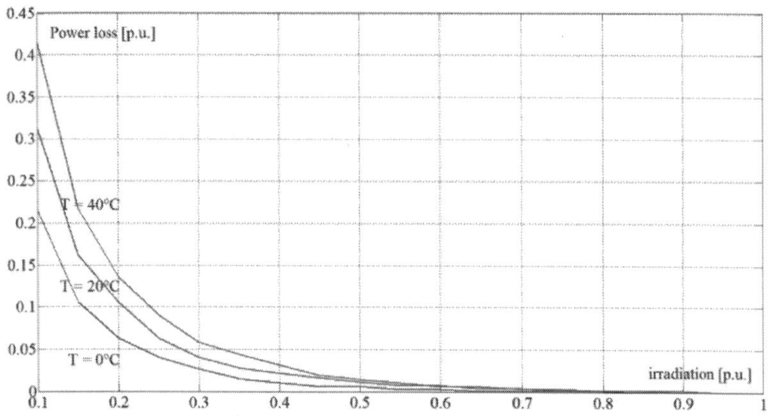

Figure 6. Power loss in p.u. *vs.* irradiation for different temperatures.

If the irradiation is higher than 0.3 p.u., the power losses are lower than 10% over a wide temperature variation range. Thus, the proposed algorithm has efficiency higher than 90% during quick transients. This error is reduced by the P&O action and is very low at a steady-state for a low value of ΔV_{min}.

6. TEST BENCH AND EXPERIMENTAL RESULTS

The MPPT technique proposed in this paper has been experimentally tested on a laboratory prototype of a PV system, shown in Figure 7. Figure 8 and Figure 9 show also a close-up view of the DIIPPC used for experiments. The test bench permits the simulation of different operating conditions of solar irradiations and temperatures.

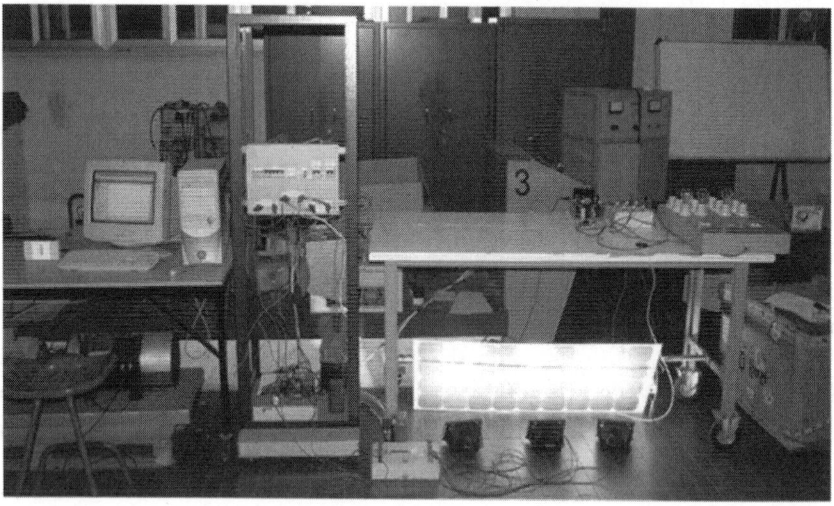

Figure 7. Photograph of the test bench used for the experimentation.

The ratings of the DIIPPC are reported in Table 3. The ripple of the current is below 1 A under rated conditions. Snubber circuits, with capacitance and resistance of 47 μF and 220 Ω, respectively, are used to protect the power Mosfets 170N10P, manufactured by IXYS (Milpitas, CA, USA). These switches have a nominal voltage of 100 V and a nominal current of 170 A. The capacitance value has been selected to store an amount of energy double than that stored in one input inductor under rated conditions. The resistance was selected on the basis of the energy discharged by the snubber capacitor at the rated voltage.

The control of the converter is based on the PIC 18F4550 microcontroller by Microchip (Chandler, AZ, USA), which features a PWM module for the gate drivers of the switches. The microcontroller inputs are the array temperature measured by a thermocouple, the PV voltage and the PV current. The thermocouple is placed on the back of the PV array, whereas the current is measured using a shunt resistor.

The solar irradiation has been simulated by means of three halogen lamps, each one having a power of 500 W. Different solar conditions are simulated regulating the power irradiated by the lamps. The use of halogen lamps is not ideal for testing PV systems. Anyway, the use of artificial light is the only method for ensuring the same variable conditions for testing different MPPT algorithms. Moreover, being the same the irradiation change, also if it is not the ideal light for optimizing the panels, the results, in terms of comparison between different MPPTs can be considered correct. That's why, also in [30] the authors use halogen lamps for simulating controllable irradiation conditions.

In order to evidence the differences between the different algorithms, some step changes in the solar irradiation have been simulated. Also if this condition is not so frequent, in nature, the step change in solar irradiation represents a good test for comparing the dynamic response of different MPPT algorithms [27].

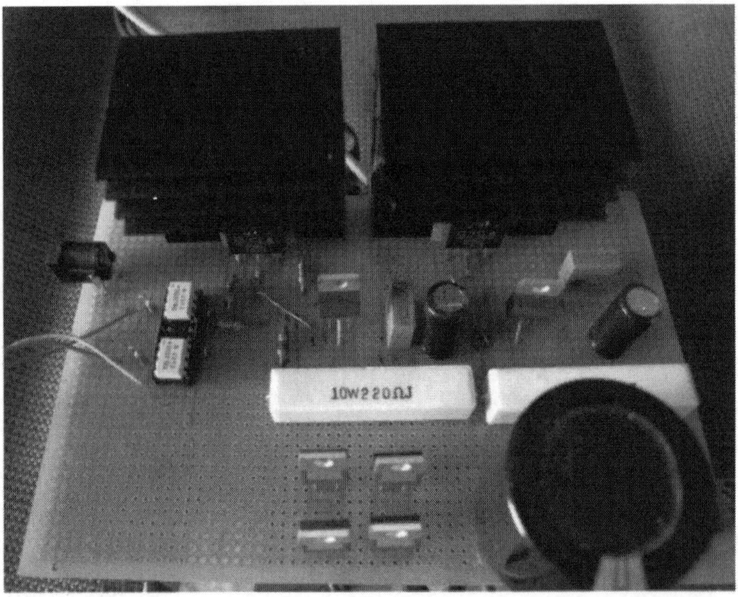

Figure 8. Close-up front view of the DIIPPC.

Table 3. Characteristics of the DC-DC converter.

Rated Power	Rated Input Voltage	Rated Output Voltage	Switching Frequency	Input Current Ripple	Drain to Source Mosfet Voltage	Inductances Value	Transformer Ratio
300 W	10 V	100 V	100 kHz	<3%	100 V	80 μH	4:12

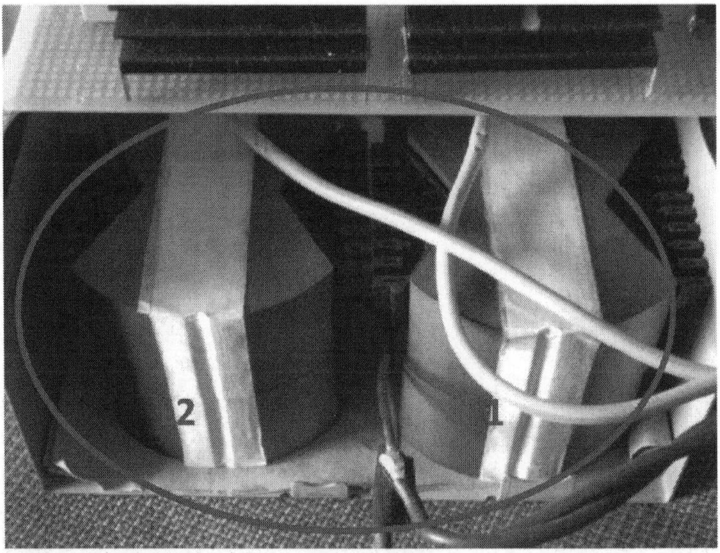

Figure 9. Close-up rear view of the DIIPPC.

A 100 Ω resistor has been used as a load. The actual load seen by PV array is dependent on the duty-cycle of the converter according to the following relationship:

$$R_{actual} = R_{load} \frac{(1-D)^2 N_1^2}{N_2^2} \quad (17)$$

A block diagram of the system control is represented in Figure 10. The MPPT algorithm is implemented by changing the duty-cycle of the DIIPPC via a proportional-integral regulator. The constants of the regulator have been selected in order to obtain a good compromise between good dynamic of MPPT and reduced steady-state oscillations around the MPP. After several tests, have been chosen the values: $k_p = -0.03$ and $k_i = 1.5$. The two constants are negative because, for the electrical characteristic of the PV source, an increasing duty cycle of the converter implies a reduction of the output voltage of the PV system.

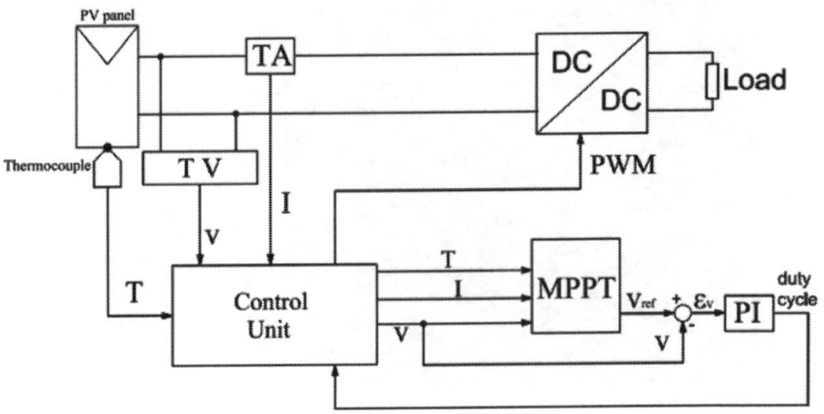

Figure 10. Block diagram of the control system.

The modified P&O algorithm proposed in this paper has been experimentally compared to a traditional P&O. In order to enhance the performances of the traditional P&O, some preliminary tests were carried out in order to find the best perturbation step, *i.e.* the step capable of extracting the maximum average power from the PV panel. To this aim, a sample irradiation pattern was created, consisting of 30 s of full irradiation, 30 s of 40% irradiation, and another 30 s of full irradiation. At a temperature of 23 °C, the best step for the traditional P&O is 0.3 V, whereas at 32 °C it is 0.7 V.

The performances of the modified P&O algorithm are strongly dependent on the function *f*, which defines the adaptive step of the algorithm. Experimental tests showed that the best results are obtained when *f* is a quadratic polynomial function, with a minimum perturbation step of 0.05 V at the maximum power point voltage under rated conditions. This result is in accordance with the consideration reported in Section 3.

The results of the comparison are presented in Figure 11 for a temperature of 23 °C. The left side of Figure 11 clearly shows that the proposed modified P&O algorithm is better than the standard P&O at steady-state, because the oscillations are smaller and the average power level is higher. During transients (starting at $t = 30$ s and $t = 60$ s) the proposed method shows the same reactivity of the standard P&O, as shown in the right side of Figure 11.

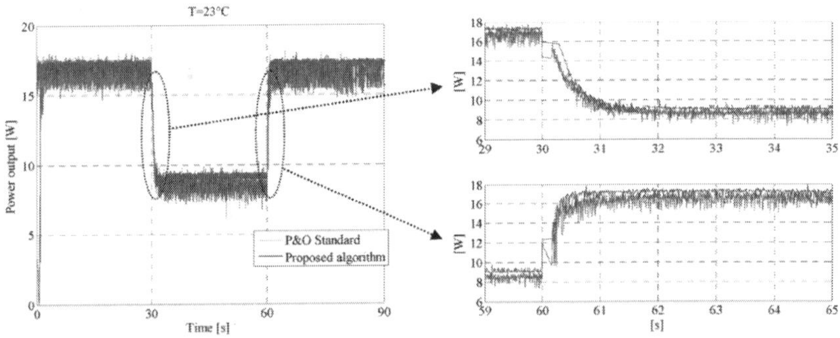

Figure 11. Comparison between the proposed algorithm and a standard P&O at 23 °C.

This means that the enhanced oscillations around the maximum of the proposed method is not paid with a slower dynamic capability, as it would happen for a traditional P&O algorithm. Indeed the perturbation step is higher when the irradiation changes suddenly and is lower when the operating point is close to the maximum power point. Moreover, the good performance at steady-state is also guaranteed by the smooth variation of the perturbation step, which is obtained by the quadratic shape of the function *f*. Figure 12 shows a similar behaviour for a temperature of 32 °C.

It is important to notice that at 32 °C the perturbation voltage step for the traditional P&O has been changed from 0.3 to 0.7 in order to assure the maximum average power and a good dynamic; this adjustment was not necessary for the proposed modified P&O technique, which shows the same reactivity for different temperatures. The proposed MPPT algorithm has also been compared to the IC algorithm. As for the case of the traditional P&O, the IC algorithm was tuned in order to define the best perturbation step. The comparison between the proposed algorithm and IC algorithm is reported in Figure 13 for a temperature of 23 °C.

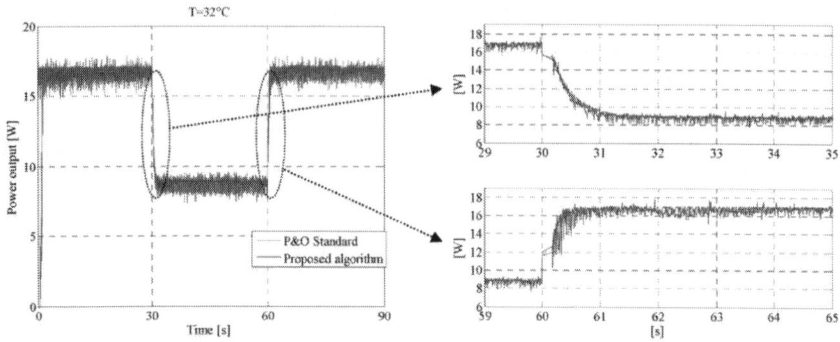

Figure 12. Comparison between the proposed algorithm and a standard P&O at 32 °C.

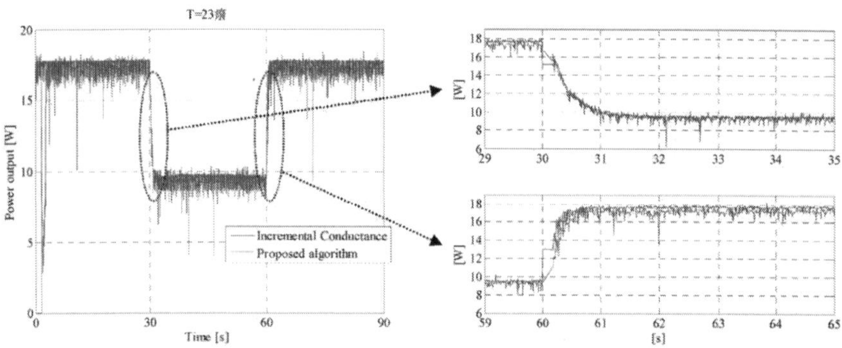

Figure 13. Comparison between the proposed algorithm and IC method at 23 °C.

The diagrams show that situation is similar to that of Figure 11, where the proposed algorithm shows smaller oscillations and higher average power level together with comparable dynamics during transients. Figure 14 shows that a similar behaviour is observed when the temperature is 32 °C. Finally Table 4 summarizes the average power experimentally obtained from the panel using the three algorithms under the solar irradiations of Figure 10, Figure 11, Figure 12, Figure 13and Figure 14.

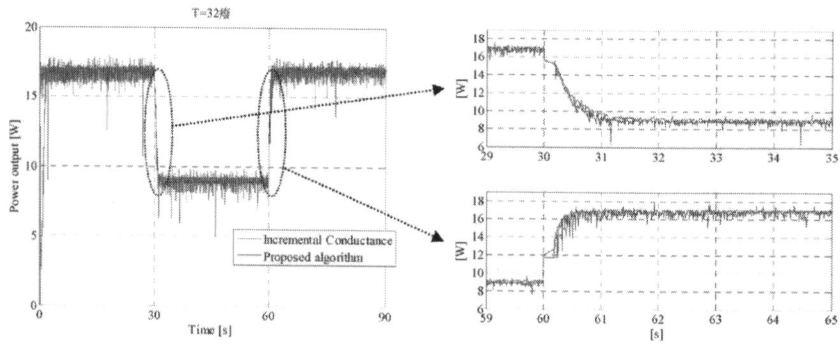

Figure 14. Comparison between the proposed algorithm and IC method at 32 °C.

Table 4. Average power obtained from different algorithms.

Temperature	Standard P&O	Incremental Conductance	Modified P&O
23 °C	14.1 W	14.1 W	15.0 W
32 °C	14.1 W	14.1 W	14.5 W

As it can be seen from the values of Table 4, the proposed algorithm reaches higher average power in comparison to traditional methods; the increment is 3.6% depending on the operating temperature.

An extensive study was carried out in order to evaluate the conversion efficiencies of the three algorithms considered in this paper. For each operating temperature, the solar irradiation was estimated by the short circuit current, which is directly proportional to the irradiation. For the nominal irradiation of 1000 W/m^2 and temperature of 25 °C, the short circuit current is about 300 A/m^2. For different temperatures, the short circuit current has been modified by the factor 0.2%/°C, as explained in Section 4. Therefore, the power reaching the panel is the product of the instantaneous irradiation times the area of the array A_c. If the power output of the panel is indicated as p(t) and the solar irradiation as $I(t)$, the efficiency of the panel, η, can be evaluated as follows:

$$\eta = \frac{\int_0^T p(t)dt}{A_c \int_0^T I(t)dt} \quad (18)$$

Table 5 gives the results of the comparison of the three algorithms for several operating temperatures in terms of panel efficiencies. From the examination of Table 4 and Table 5, it is clear that the modified P&O algorithm proposed in this paper has efficiency higher than that of the algorithms traditionally described in the technical literature. The shape of the smoothing function, which modifies the perturbation step, plays a key role in improving the transitions after a change in the solar irradiation and reduces the steady-state oscillations around the MPP.

Table 5. Panel efficiencies obtained from different algorithms.

Temperature	23 °C	26 °C	28 °C	30 °C	32 °C	35 °C	36 °C	40 °C	44 °C
Standard P&O	8.9%	9.0%	9.2%	9.0%	8.9%	9.0%	9.0%	8.8%	8.4%
Incremental Conductance	9.1%	9.2%	8.8%	9.1%	8.9%	9.0%	9.1%	8.5%	8.5%
Modified P&O	9.5%	9.6%	9.3%	9.3%	9.2%	9.2%	9.2%	8.9%	8.6%

7. ADDITIONAL DISCUSSION ON THE MPP ESTIMATOR ERROR

As additional investigations a numerical analysis is carried out supposing that the actual MMP is estimated with a ±10% of error; that produces a ΔV_{min} according to Equation (15). In these simulations the irradiation starts from value of 1 p.u., at time of 0.5 sec. decrease to 0.45 p.u. and gets back to 1 p.u. at time of 1.0 sec. For the proposed modified P&O algorithm,Figure 15 shows the output power in the cases $V_{M,est} =$ 0.9 V_M and $V_{M,est} =$ 1.1 V_M, together with the perfect estimation case $V_{M,est} = V_M$. Since the performance of a traditional P&O method is strongly influenced by the voltage perturbation step, several simulations were carried out. In Figure 16 the result of a traditional P&O algorithm with the settings $\Delta V = \Delta V_{min}$, $\Delta V = 2\ \Delta V_{min}$ and $\Delta V = 4\ \Delta V_{min}$, is reported in terms of output power. The simulations shows that, for the irradiation pattern selected, the traditional P&O method is unable to reach the MPP when $\Delta V = \Delta V_{min}$. Therefore the minimum step has to be at least $2 \cdot \Delta V_{min}$, with consequent reduction of the steady-state performance of the MPP algorithm. Thus, a setting of $2 \cdot \Delta V_{min}$has been chosen to make comparison with the proposed method; for the proposed method the setting $V_{M,est} = 0.9\ V_M$ is chosen, corresponding to the worst case (minimum mean output power).

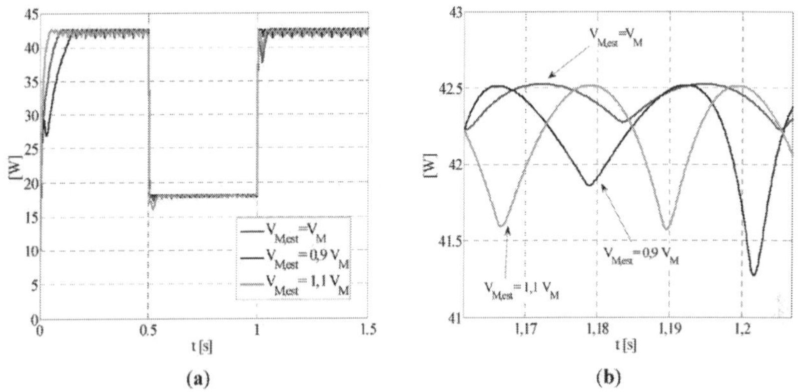

Figure 15. Output power when using the proposed method for three different estimation errors; complete simulation (**a**) and steady-state analysis (**b**).

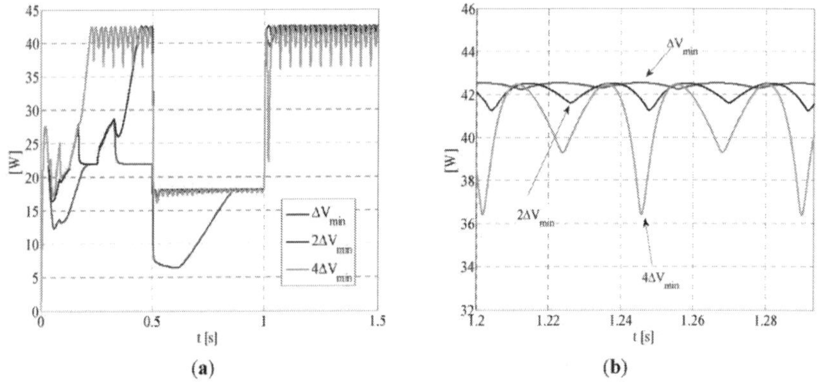

Figure 16. Output power when using the standard P&O method for three different perturbation steps; complete simulation (**a**) and steady-state analysis (**b**).

In Figure 17 the best result of traditional P&O algorithm ($\Delta V = 2\ \Delta V_{min}$) is compared to the worst case of the proposed algorithm ($V_{M,est} = 0.9\ V_M$).

From Figure 17 it can be deduced that the proposed algorithm is faster than the traditional P&O method during transient and ensures smaller oscillation in steady-state conditions. As a consequence, the proposed method produces a higher mean power value, even when an estimation MPP error of 10% is introduced.

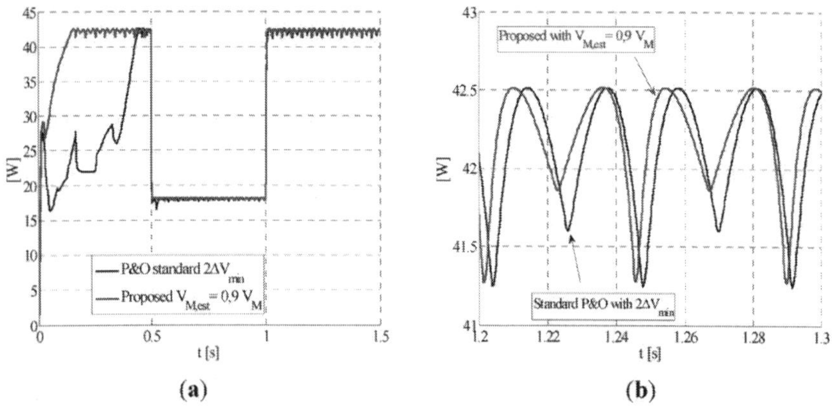

Figure 17. Comparison of output power between traditional P&O and proposed method with 10% estimation error; complete simulation (**a**) and steady-state analysis (**b**).

8. CONCLUSIONS

A variation of the classic maximum power point tracking perturb & observe method for photovoltaic power systems has been proposed, implemented and tested. The main limitation of perturb & observe algorithms is the compromise between fast dynamic response and steady-state stability. To improve the dynamic performance and ensure good stability, a criterion for choosing the optimal perturbation step and its coefficients has been discussed. The chosen function enables more flexibility of algorithm to the actual photovoltaic system conditions.

The modified perturb & observe algorithm has been implemented on a dual input inductor push-pull converter and experimentally compared to traditional P&O and IC techniques. For both these algorithms, the perturbation step was adjusted with tuning tests in order to extract the maximum average power from the PV panel and, consequently, penalizing the proposed system in the comparison.

The experimental results have demonstrated clearly that the proposed modified perturb & observe algorithm is better than the standard P&O and IC at steady-state and shows comparable reactivity during transients, even if no adjustment is operated on the perturbation voltage step for different operating conditions. This is due to the intrinsic adaptive characteristic of the algorithm: the perturbation step is higher when the irradiation changes suddenly and is lower when the operating point is close to the maximum power point. From the analysis of the results, it is evident that the proposed perturb & observe algorithm has higher efficiencies than the algorithms already proposed in the technical literature.

AUTHOR CONTRIBUTIONS

The authors participated at the paper through a wide and balanced cooperation. They all worked at the mathematical model, the simulation set-up, the experiments and writing the manuscript. They have given equal contribution regarding all aspects of the paper.

CONFLICTS OF INTEREST

The authors declare no conflict of interest.

REFERENCES

1 Esram, T.; Chapman, P.L. Comparison of photovoltaic array maximum power point tracking techniques. *IEEE Trans. Energy Convers.* **2007**, *22*, 439–449.

2 Liu, X.; Lopes, L.A.C. An improved perturbation and observation maximum power point tracking algorithm for PV arrays. In Proceedings of the Power Electronics Specialists Conference PESC, Aachen, Germany, 20–25 June 2004; pp. 2005–2010.

3 Kim, I.S.; Kim, M.B.; Youn, M.J. New maximum power point tracker using sliding-mode observer for estimation of solar array current in the grid-connected photovoltaic system. *IEEE Trans. Ind. Electron.* **2006**, *53*, 1027–1035.

4 Cabal, C.; Alonso, C.; Cid-Pastor, A.; Estibals, B.; Seguier, L.; Leyva, R.; Schweitz, G.; Alzieu, J. Adaptive digital MPPT control for photovoltaic applications. In Proceedings of the ISIE, Vigo, Spain, 4–7 June 2007; pp. 2414–2419.

5 Sera, D.; Kerekes, T.; Teodorescu, R.; Blaabjerg, F. Improved MPPT method for rapidly changing environmental conditions. In Proceedings of the IEEE, Montreal, QC, Canada, 9–13 July 2006; pp. 1420–1425.

6 Femia, N.; Petrone, G.; Spagnuolo, G.; Vitelli, M. Optimization of perturb and observe maximum power point tracking method. *IEEE Trans. Power Electron.* **2005**, *20*, 963–973.

7 Wu, T.; Chang, C.; Chen, Y. A fuzzy-logic-controlled single stage converter for PV-powered lighting system applications. *IEEE Trans. Ind. Electron.* **2000**, *47*, 287–296.

8 Khaehintung, N.; Wiangtong, T.; Sirsuk, P. FPGA Implementation of MPPT using variable step-size P&O algorithm for PV applications. In Proceedings of the ISCIT, Bangkok, Thailand, 20 September–18 October 2006; pp. 212–215.

9 Femia, N.; Granozio, D.; Petrone, G.; Spagnuolo, G.; Vitelli, M. Predictive and adaptive MPPT perturb and observe method. *IEEE Trans. Aerosp. Electron. Syst.* **2007**, *43*, 934–950.

10 Piegari, L.; Rizzo, R. Adaptive perturb and observe algorithm for photovoltaic maximum power point tracking. *IET Renew. Power Gener.* **2010**, *4*, 317–328.

11 Abdelsalam, A.K.; Massoud, A.M.; Ahmed, S.; Enjeti, P.N. High-performance adaptive perturb and observe MPPT technique for photovoltaic-based microgrids. *IEEE Trans. Power Electron.* **2011**, *26*, 1010–1021.

12 Al-Amoudi, A.; Zhang, L. Optimal control of a grid-connected PV system for maximum power point tracking and unity power factor. In Proceedings of the 7th International Conference Power Electronics and Variable Speed Drives, London, UK, 21–23 September 1998; pp. 80–85.

13 Zhang, L.; Al-Amoudi, A.; Bai, Y. Real-time maximum power point tracking for grid-connected photovoltaic systems. In Proceedings of the 8th International Conference Power Electronics and Variable Speed Drives, London, UK, 18–19 September 2000; pp. 124–129.

14 Patel, H.; Agarwal, V. MPPT scheme for a PV-fed single-phase single stage grid-connected inverter operating in CCM with only one current sensor. *IEEE Trans. Energy Convers.* **2009**, *24*, 256–263.

15 Chiang, M.-L.; Hua, C.-C.; Lin, J.-R. Direct power control for distributed PV power system. In Proceedings of the Power Conversation Conference, Osaka, Japan, 2–5 April 2002; pp. 311–315.

16 Pandey, A.; Dasgupta, N.; Mukerjee, A.K. High-performance algorithms for drift avoidance and fast tracking in solar MPPT system. *IEEE Trans. Energy Convers.* **2008**, *23*, 681–689.

17 Xiao, W.; Dunford, W.G. A modified adaptive hill climbing MPPT method for photovoltaic power systems. In Proceedings of the IEEE 35th Annual PESC, Aachen, Germany, 20–25 June 2004; pp. 1957–1963.

18 Wolfs, P.J.; Tang, L. A single cell maximum power point tracking converter without a current sensor for high performance vehicle solar arrays. In Proceedings of the IEEE 36th PESC, Recife, Brazil, 16 June 2005; pp. 165–171.

19 Jain, S.; Agarwal, V. A new algorithm for rapid tracking of approximate maximum power point in photovoltaic systems. *IEEE Power Electron. Lett.* **2004**, *2*, 16–19.

20 D'Souza, N.S.; Lopes, L.A.C.; Liu, X. An intelligent maximum power point tracker using peak current control. In Proceedings of the IEEE 36th PESC, Recife, Brazil, 16 June 2005; pp. 172–177.

21 Manna, B.; Saha, N.; Palit, D. Optimization of solar power using servo mechanism by Perturb & Observe method. In Proceedings of the 2014 Annual IEEE India Conference (INDICON), Pune, India, 11–13 December 2014; pp. 1–6.

22 Femia, N.; Petrone, G.; Spagnuolo, G.; Vitelli, M. A technique for improving P&O MPPT performances of double-stage grid-connected photovoltaic systems. *IEEE Trans. Ind. Electron.* **2009**, *56*, 4473–4482.

23 Desai, H.P.; Patel, H.K. Maximum power point algorithm in PV generation: An overview. In Proceedings of the 7th International Conference Power Electron Drive System, Bangkok, Thailand, 27–30 November 2007; pp. 624–630.

24 De Aragao Filho, W.C.; Barbi, P.I. A comparison between two current-fed push-pull DC-DC converters-analysis, design and experimentation. In Proceedings of the 18th INTELEC '96, Boston, MA, USA, 6–10 October 1996; pp. 313–320.

25 Nousiainen, L.; Puukko, J.; Mäki, A.; Messo, T.; Huusari, J.; Jokipii, J.; Viinamäki, J.; Lobera, D.T.; Valkealahti, S.; Suntio, T. Photovoltaic generator as an input source for power electronic converters. *IEEE Trans. Power Electron.* **2013**,*28*, 3028–3038.

26 Suntio, T.; Leppäaho, J.; Huusari, J.; Nousiainen, L. Issues on solar-generator interfacing with current-fed MPP-tracking converters. *IEEE Trans. Power Electron.* **2010**, *25*, 2409–2419.

27 Sera, D.; Teodorescu, R.; Hantschel, J.; Knoll, M. Optimized maximum power point tracker for fast changing environmental conditions. In Proceedings of the IEEE International Symposium

on Industrial Electronics, Cambridge, UK, 30 June 2008–2 July 2008; pp. 2401–2407.

28 Khanaki, R.; Radzi, M.A.M.; Marhaban, M.H. Comparison of ANN and P&O MPPT methods for PV Applications under Changing Solar Irradiation. In Proceedings of the 2013 IEEE Conference on Clean Energy and Technology (CEAT), Lankgkawi, Malesia, 18–20 November 2013; pp. 287–292.

29 Zorrilla-Casanova, J.; Piliougine, M.; Carretero, J.; Bernaola, P.; Carpena, P.; Mora-López, L.; Sidrach-de-Cardona, M. Analysis of dust losses in photovoltaic modules. In Proceedings of the World Renewable Energy Congress, Linköping, Sweden, 8–13 May 2011; pp. 2985–2992.

30 Galan, D.M.; Starzak, L.; Torzewicz, T.; Piotrowicz, M.; Maranda, W. Laboratory setup for investigation of MPPT algorithms of photovoltaic modules under non-uniform insolation. In Proceedings of the 20th International Conference Mixed Design of Integrated Circuits and Systems (MIXDES), Gdynia, Polonia, 20–22 June 2013; pp. 603–606.

CHAPTER 10

Power Quality Issues Concerning Photovoltaic Generation and Electrical Vehicle Loads in Distribution Grids

Jaan Niitsoo[1], Paul Taklaja[1], Ivo Palu[1], Joni Klüss[2]

[1]Department of Electrical Power Engineering, Tallinn University of Technology, Tallinn, Estonia

[2]Department of Electrical Engineering and Automation, Aalto University, Helsinki, Finland

ABSTRACT

The high utilization level of renewable generation including residential photovoltaic (PV) systems together with the uncontrolled charging of electric vehicles (EVs) can have a significant impact on load characteristics in distribution networks. Harmonic content of PV generation, EV charging loads, and their influence on power quality indicators in residential distribution networks are discussed in this paper. For investigating likely power quality scenarios, PV generation and EV charging measurement results including current harmonic amplitude and phase angle values are used and compared with present load characteristics. Different modelling scenarios are analysed and a simplified model of harmonics in PVs and EVs is offered. The results of the study show moderate additional harmonic distortion in residential load current and voltage distortion at the substation's busbar when PV generation and EV loading are added. The scenarios presented in this paper can be further used for modelling the actual harmonic loads of the PVs and EVs in distribution networks.

KEYWORDS

Current Measurement, Distributed Generation, Electric Vehicles, Load Modelling, Photo Voltaic, Power Quality, Power System Harmonics

1. INTRODUCTION

Power distribution networks are designed to operate with sinusoidal voltage. The deviation of the actual voltage waveform from an ideal sine wave is a fundamental aspect related to power quality. This deviation can be described as a sum of harmonic voltage components. A higher level of harmonics correlates to larger distortion. From the point of view of electromagnetic compatibility, harmonics must be kept within given compatibility levels to enable satisfactory operation of all equipment supplied by the network. Furthermore, since electricity is also defined as a product, utility companies could be held responsible for excessively high harmonic levels and any resulting damage to customers' property [1] .

The appliances and electric devices connected to the public power supply network are also designed to operate with a sinusoidal voltage at rated power. However, many of the connected loads are nonlinear, meaning that they draw current with a distorted sine waveform. It has been estimated that already in 2012, 60% of the power system loads in USA were nonlinear loads [2] . Nonlinear current in turn provides nonsinusoidal voltage drop, thus resulting in distorted network voltage.

In particular, distortion levels may increase significantly when numerous harmonic emitting devices are utilized in bulk [3] .

Distorted voltage and current may result in undesirable effects not only for the distribution system, but also for the customers. Nonlinear loads inject harmonic currents and induce increased voltage drops over both phase and neutral conductors [4] . As a result, potential damage to sensitive electronic equipment and capacitor banks, overheating of transformers and neutral conductors, and additional losses in the power system are likely. Degraded power quality entails additional costs for both the electricity distributor and its customers as these effects cause premature aging and failure in power system devices [5] - [7] .

Advances in power electronics has led to the widespread use of switching converters for the general public use as well as industrial applications. Power electronic converters draw non-sinusoidal current from the grid and this has led to a rapid increase in the number of nonlinear loads. In addition to the increased number of electronic devices, also resistive devices such as incandescent lamps are ever more frequently replaced by energy saving lamps utilizing nonlinear elements. Depending on type and brand, switching power supplies absorb distorted currents which flow through the impedances of the power distribution system and result in distortion of system bus voltage [8] .

Similarly, modern EV chargers employ switching power converters. Due to the high energy requirement of the EV, charging at home can use power levels very close to the ones of the actual residential customers drawing, 10 - 15 A [9] . The charging of EVs produces considerable additional power electronic load which can generate harmonics and result in associated power quality issues in residential networks.

Power electronics is also implemented in residential photovoltaic (PV) generators, which is currently the dominant renewable energy source in urban and metropolitan areas. The inverters required to supply households from the PVs are nonlinear power supplies, meaning that they provide distorted waveform outputs that increase the levels of harmonics. This technology is enjoying rapid growth due to a combination of subsidies, the abundance of sunshine, and the low impact of the technology on the urban landscape [5] .

The present study identifies and analyses possible power quality scenarios in a residential distribution network by examining the impact of nonlinear domestic loads such as EVs and PV inverters. Measured power consumption and current waveforms of different home appliances, EVs and PV inverter have been used for the analysis. All loads considered in the model were provided with magnitudes and phase angles of each harmonic up to the 50th order. The main purpose of this paper is to present the use of actual measurement data from different devices for modelling the effects on the residential distribution network and give an estimation of the important values for further modelling.

2. THEORETICAL BACKGROUND

In electrical power networks, a non-sinusoidal periodic waveforms can be presented as a sum of numerous harmonic components (harmonics), each having an integer-multiple frequency of the main frequency. Different waveforms have different harmonic content, referring to different patterns of individual harmonic magnitudes and phase shift compared to the main frequency component. Hereafter in this paper, the presented measurements of loads are all indicated as magnitudes and phase shift of each individual harmonic up to the 50th order.

Distortions can be observed individually by comparing different harmonic components and calculating harmonic distortion (HD). A more general approach to quantifying the distortions is using the total harmonic distortion level (THD). Total harmonic distortion can be expressed separately for current harmonic distortion as THDI and for voltage distortion as THDU. The total harmonic distortion indicators can be calculated using corresponding Equations (1) and (2),

$$\mathrm{THD_I} = \frac{1}{i_1}\sqrt{i_{rms}^2 - i_1^2} \times 100 \tag{1}$$

$$\mathrm{THD_U} = \frac{1}{u_1}\sqrt{u_{rms}^2 - u_1^2} \times 100 \tag{2}$$

where i_1 is current of 1st order and u_1 is voltage of 1st order.

THD does not reveal the magnitudes of individual harmonics, which could still exceed the limits for specific harmonics regardless of THD value. For the correct estimation of the harmonic levels, calculations have to make use of magnitude and phase angle values of individual harmonics. Harmonic phase angle diversity is relevant when multiple appliances are operating simultaneously, creating either reinforcement or cancellation of harmonic magnitudes [2] . It is reported that 10% smaller harmonic current magnitudes can be seen when phase angle information is included compared to the simple summing of magnitudes without phase angle values [10] . The attenuation effect is dependent only on the phase angle, but the effect's severity is dependent on the magnitude of the harmonic voltage [11] .

The combined influence of harmonics from different sources is highly dependent on network topology, mutual conductor impedances and phase balance. The resulting harmonic distortion in the distribution network can increase or decrease due to the variation of phase angles for different harmonic sources. A special situation occurs in the case of a balanced three phase load consisting of identical nonlinear loads. Since the triplen harmonics sum in neutral, the three phase supply system neutral conductor total current can be higher than the phase current [4] [12] . For example in the Netherlands and in Denmark, analysis has shown that the 15th harmonic current, which is one of the triplen harmonics, has exceeded the tolerable limits in several cases [12] . Therefore, phase balance is a significant factor for harmonic emission in distribution networks [7] .

The harmonic generation of a PV system depends on the inverter technology, and operation conditions, including solar irradiance, temperature, loads, and the supply system characteristics. Both the current THD and the output reactive power are related to the output active power levels, which in turn are strongly dependent on solar irradiance levels. Most of the inverters consume or feed reactive power into the network depending on their output active power and their technology. During operation at low solar irradiance levels (e.g. sunrise, sunset, cloudy days), current THD values can increase rapidly since the THD factor is inversely proportional to the output active power of the PV inverters. Nevertheless, THD is notably reduced as the output active power of the PV inverters increases and reaches its nominal value [13] - [15] .

Varying power density of renewable energy not only potentially cause supply voltage sag or swell patterns but also frequency variations in the LV grids. The application of power converters as interfaces between energy sources and the LV grid and their interaction with other system components may cause high harmonics distortion [16] [17] . The effects of the nonlinear residential load on voltage THD are most significant on a local level, and equipment failures due to voltage distortion are more likely to occur along distribution feeders than farther upstream. The results show that recommended voltage THD limits can easily be exceeded, particularly at nearby distribution feeder tap points, if the loading equipment is highly nonlinear [18] .

EV batteries require DC for charging, but all uncontrolled rectifiers inject a high content of current harmonics into AC power networks. Based on literature, it can be concluded that over time and with the development of power supply technologies, the amount of distortion from EV charging has decreased. Measurements in the 1990's showed that the use of uncontrolled or low-control rectifiers caused average current THD of 50% [19] . Measurements of modern commercial EVs have indicated an average charging current THD around 11%...12% [20] [21] , but values as low as 4.5% [22] have also been indicated.

Monitoring of 100 distribution network feeders in USA revealed that the average voltage THD at PCC (Place of Common Coupling) was 4.73% [23] . It may be assumed that the stated percentage is quite common for many distribution grids. Power supply standards (EN50160) have set the upper voltage THD limit for public networks at 8% [24] . Nevertheless, severe cases have been revealed where distortion has been higher than 17% [25] .

3. METHODS

The 50 Hz residential distribution network at 0.4 kV and loads for assessing load flow were modelled using DIgSILENT Power Factory software. The model consisted of a three-phase residential load at 0.4 kV voltage level composed of different single phase loads. The schematic of the residential load model is presented in Figure 1.

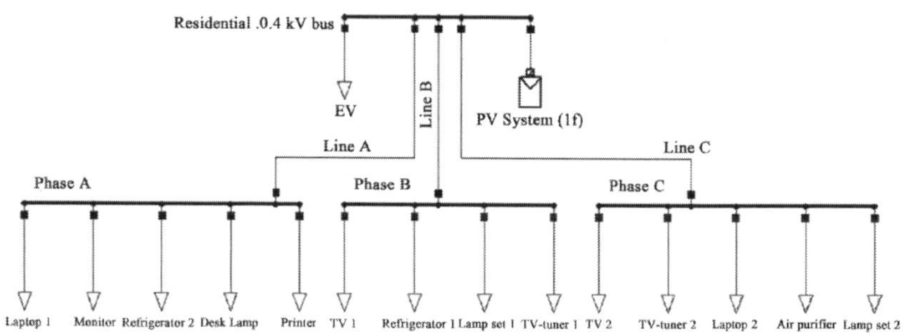

Figure 1. Schematic of residential load model.

The compiled residential load was connected to the distribution network substation via a 1.4 km long 4 × 95 mm^2 overhead line (OHL) as depicted on Figure 2 and it is the only consumer of this substation. The distribution network substation was connected to a 10 kV network with short-circuit power of 200 MVA and short-cir- cuit current 11.5 kA. The high voltage (HV) busbar is modelled as a slack bus. The transformer used in the distribution substation was modelled with the following parameters:

Ÿ nominal power 25 kVA;

Ÿ relative short circuit voltage 4.5%;

Ÿ zero sequence impedances $r_0 = 0.02$ pu and $x_0 = 0.04$ pu;

Ÿ magnetizing impedance/short circuit impedance ratio 3;

Ÿ vector group Yyn.

Implemented parameters in the simulation were selected based on power quality problematic issues identified in Elektrilevi's network (Estonia's main distribution grid operator) for July 1, 2013. The length of the OHL between substation and customer's PCC was defined as an average of all the lines between substations and customers with power quality problems. Similarly, the selected diameter of the line and nominal power of the transformer are the most common values for the identified problematic components.

Harmonic voltage amplitudes and phase angles up to the 50th order were obtained from measurements conducted by Elektrilevi at one of the sites where power quality issues were identified. Harmonic voltage distortion at the 10 kV bus was measured and modelled around 2%, which is a common value for this grid.

For modelling PV generation, one single phase PV was measured for one week. For modelling EV loads, actual charging process of five different EVs were measured. All vehicles were produced in 2012-2013 and are commercially available on the markets. In each case, harmonic current amplitudes and phase angles up to 50th order were measured and used in the models in DIgSILENT. Both PV and EV were connected to residential load's busbar as was described in Figure 1. Also a mean load model of averaged values was composed for the single phase PV system and EV.

In order to model the network response of nonlinear loads, 14 different home appliances were measured. The results of the corresponding measured active and reactive power, harmonic current magnitudes and harmonic current phase shift angles of measured devices are presented in [26] [27] . Modelled devices were arranged in a manner where similar active power consumption was seen in every phase. In the model, all nonlinear devices are in operation and coincidence factors are not taken into account, thus representing the presumed worst case scenario.

4. RESULTS

Three cases modelling the residential distribution network are presented. Initial conditions represent the loads common to present day households.

1) First case: one single-phase PV unit connected to the PCC in addition to the initial load.

2) Second case: one EV charging load connected to the PCC in addition to the initial load.

3) Third case: EV charging load and PV inverter are connected to the PCC in addition to the initial load.

Initial values of voltage and current in the grid before adding PV generations or EV load are presented in Table 1. Voltage THD and power factor (PF) values are listed in Table 2 and power values are listed in Table 3.

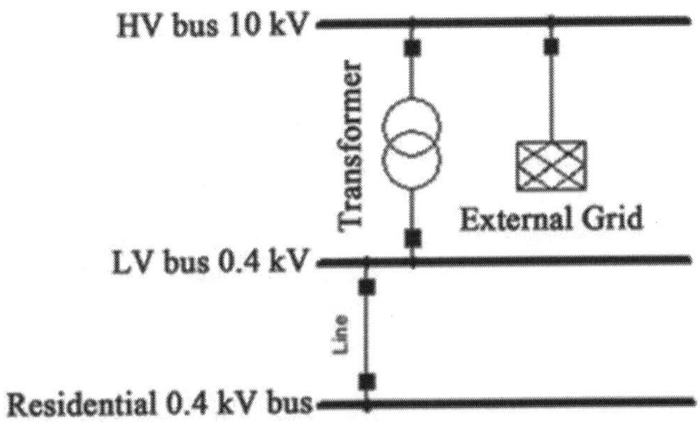

Figure 2. Schematic of distribution grid model.

A) First Case—Single Phase PV

Measurement results for three different power levels (1—near 30%, 2—near 60%, 3—near 100%) of the single phase PV are given in Table 4. From the table, it can be assumed that current distortion is slightly correlated with current and decreases as current increases. Also voltage and voltage distortion variations may somewhat have an effect on the current distortion, but are here ignored due to very low values of voltage THD and the fact that current itself is more affected by voltage. The same conclusion can be made by observing the power factor (PF) value which approaches unity with increasing current. Interestingly in this case, reactive power Q appears to be independent of the current level and changes polarity.

Voltagc and current distortion during a 15 h period is shown in Figure 3. Voltage distortion at the measurement point was notably low (around

1%) throughout the observed time period and similarly, current distortions are not greatly affected by grid disturbances.

Power quality indexes are illustrated in Figure 4. Results are in line with the prior conclusion for reactive power where Q is mainly capacitive throughout the measurement period. It can also be confirmed that reactive power is independent of current (compare with active power P and apparent power S).

As is evident in Figure 5, cos(φ) remained near unity throughout the measurement period, whereas PF varied considerably. The observed fluctuations in PF are a result of current distortion which is evident when comparing the current THD (THD_I in Figure 3) with the PF curve in Figure 5.

Table 5 and Table 6 present the harmonic currents and phase angles up to the 21st order for the single phase PV inverter with the corresponding power levels (1—near 30%, 2—near 60%, 3—near 100%) described in Table 4. Average values are also calculated and presented for modelling mean one phase PV. Even and higher order harmonics are left out due to their marginal dimension. All presented harmonic current amplitudes exhibited relatively moderate values, except for the third harmonic which was more notable.

For modelling mean PV generation, average values of the presented current harmonic amplitudes and angles (Table 5 and Table 6) were calculated. Main frequency current phase angles were defined zero as in the ideal case and other angles were calculated in relation to mains current.Figure 6 shows a graphical representation (vector quadrant) of the calculated average harmonics, where X and Y components of current are calculated using Formulas (3) and (4). Where A is mean harmonic (3rd, 5th, 7th etc. order) current in percentages to main frequency current and α is mean harmonic (3rd, 5th, 7th etc. order) current angle in degrees to main frequency current.

$$X = A \times \cos \alpha \tag{3}$$

$$Y = A \times \sin \alpha \tag{4}$$

A single phase PV inverter was connected to the residential busbar at different phases one at a time. The resulting voltages and currents are presented in Table 7. Voltage THD and PF is shown inTable 8 and power values are displayed in Table 9.

As it can be seen from Table 7 voltage in the phase where the PV is connected rises more than 5%. Significant voltage THD rise due to the connected PV could not be seen. Instead, THD decreased slightly in cases where PV was connected to phase A and C as evident in Table 8. Due to the change in reactive power, seen in Table 9, the PF value changed in the phase where the PV was installed.

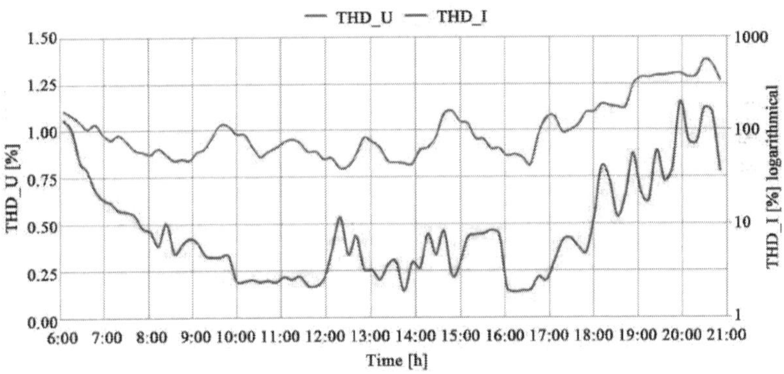

Figure 3. Measured voltage THD and current THD of single phase PV inverter.

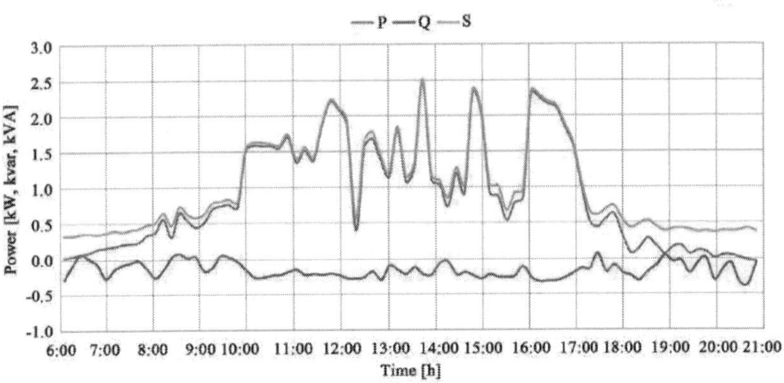

Figure 4. Measured power values of single phase PV inverter.

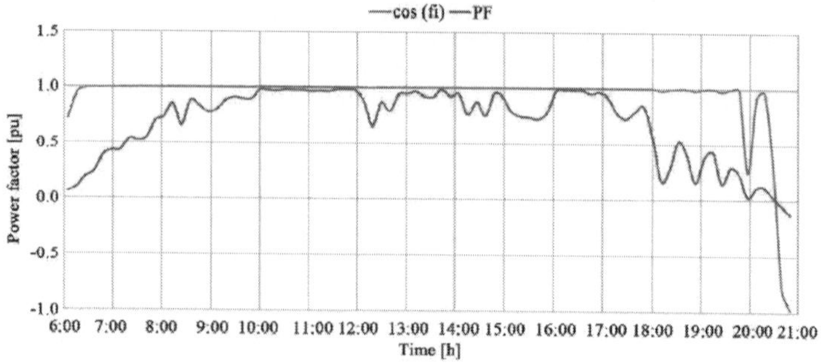

Figure 5. Measured power factors of single phase PV inverter.

Table 1. Initial modelled voltage [V] and current [A] values.

U_a	U_b	U_c	I_a	I_b	I_c	I_n
230	230	231	1.0	1.3	1.3	0.7

Table 2. Initial modelled voltage THD [%] and PF values.

THD_a	THD_b	THD_c	PF_a	PF_b	PF_c
2.9	2.1	3.2	1.00	0.92	0.84

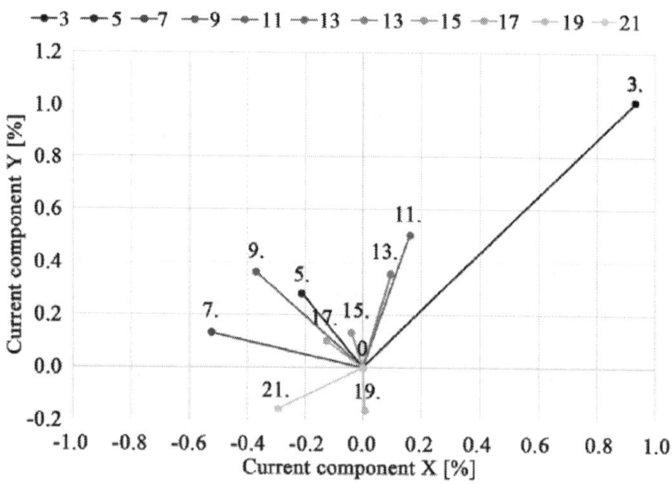

Figure 6. Mean harmonic current vectors (from 3rd to 21st order) in percentages compared to fundamental harmonic current vector (1st order) values of single phase PV inverter.

Table 3. Initial modelled active [kW] and reactive power [kvar] values.

P_a	P_b	P_c	Q_a	Q_b	Q_c
0.23	0.28	0.25	−0.01	−0.12	−0.16

Table 4. Measured values for single phase PV inverter.

Power level	1% - 30%	2% - 60%	3% - 100%
THD_U [%]	1.01	0.82	1
THD_I [%]	4.27	1.98	1.67
U_{rms} [V]	233.6	238.8	239.1
I_{rms} [A]	3.45	9.08	11.73
P [W]	739	2125	2783
Q [var]	322	−425	−257
S [VA]	807	2168	2805
cos(fi)	1	1	1
PF	0.92	0.98	0.99

Table 5. Harmonic currents of single phase PV inverter.

Order	I_1 [%]	I_2 [%]	I_3 [%]	I_mean [%]
1	100	100	100	100
3	1.92	1.2	1.01	1.38
5	0.48	0.31	0.26	0.35
7	1.08	0.27	0.27	0.54
9	0.88	0.33	0.35	0.52
11	0.89	0.38	0.32	0.53
13	0.69	0.21	0.2	0.37
15	0.23	0.08	0.1	0.14
17	0.35	0.07	0.06	0.16
19	0.29	0.1	0.1	0.16
21	0.63	0.2	0.17	0.33

Table 6. Harmonic current phase angles of single phase PV inverter.

Order	Angle_1 [°]	Angle_2 [°]	Angle_3 [°]	Angle_mean [°]
1	0	0	0	0
3	28	62	52	47
5	97	146	139	127
7	175	159	164	166
9	116	145	146	135
11	104	60	52	72
13	75	73	77	75
15	109	104	107	107
17	88	145	190	141
19	287	259	271	272
21	209	205	209	208

Table 7. Modelled voltage [V] and current [A] values for first case.

Phasing	U_a	U_b	U_c	I_a	I_b	I_c	I_n
PV in A	246	223	229	11.0	1.3	1.3	12.3
PV in B	228	245	224	1.0	10.8	1.3	11.2
PV in C	224	228	246	1.0	1.3	10.9	12.3

Table 8. Modelled voltage THD [%] and PF values for first case.

Phasing	THD_a	THD_b	THD_c	PF_a	PF_b	PF_c
PV in A	2.7	2.2	3.2	−1	0.91	0.85
PV in B	3	2.1	3.3	1	−1	0.83
PV in C	3	2.1	3	1	0.93	−1

Table 9. Modelled active [kW] and reactive power [kvar] values for first case.

Phasing	P_a	P_b	P_c	Q_a	Q_b	Q_c
PV in A	−2.58	0.28	0.25	0.23	−0.13	−0.16
PV in B	0.23	−2.54	0.25	−0.01	0.13	−0.17
PV in C	0.24	0.28	−2.56	−0.02	−0.11	0.09

B) Second Case—EV

Measurements were recorded for the charging of five different EVs. Power and THD values while EVs were charging at constant power level are presented in Table 10. As the power values (P and Q) were quite similar for all five EVs, the current distortion varied more. Three EVs out of five had current distortion around 10%...12% and two had substantially lower values, around 3%...4%.

The characteristics of individual current harmonic magnitudes and phase angles for odd harmonics during the constant power charging are presented in Table 11 and Table 12. The highest current amplitudes were observed at 3rd, 5th, 7th and 13th orders. For successful modelling, the mean values based on five EVs were calculated. The arithmetic mean of amplitudes (AM) and geometric mean for phase angles (GM) was calculated as follows:

Table 10. General data for EVs during constant power charging time.

EV	P [kW]	Q [kVAr]	THD_I [%]
1	2.2	0.2	4.2
2	2.4	0.5	12.3
3	2.9	0.2	3.4
4	2.4	0.1	10.5
5	2.2	0.4	11.2
Mean	2.5	0.3	8.1

Table 11. Harmonic current amplitudes [%] of EVs.

Order	EV 1	EV 2	EV 3	EV 4	EV 5	Arithmetic mean
1	100	100	100	100	100	100
3	8.7	11.9	3.3	2.6	11.1	7.5
5	3.7	0.4	0.9	2	2.3	1.9
7	3.1	0.6	0.9	1.1	1.4	1.4
9	1.2	0.6	0.2	0.5	1.5	0.8
11	0.8	1.1	0.9	0.7	1	0.9
13	1.7	1.1	1.3	0.9	0.8	1.2
15	0.4	0.6	0.3	0.3	0.9	0.5
17	0.4	0.5	0.1	0.1	0.2	0.3
19	0.7	0.3	0.1	0.3	0.2	0.3
21	0.8	0.3	0.4	0.9	0.4	0.6

Table 12. Harmonic current phase angles [°] of EVs.

Order	EV 1	EV 2	EV 3	EV 4	EV 5	Geometric mean
1	0	0	0	0	0	0
3	8	33	158	327	6	39
5	157	180	335	338	294	248
7	255	269	205	213	299	245
9	46	120	287	320	126	145
11	326	256	20	67	221	120
13	30	305	298	334	26	119
15	257	199	30	194	142	133
17	207	162	246	247	291	226
19	40	168	313	55	219	121
21	154	304	165	217	166	195

$$AM = \frac{1}{n}(a_1 + a_2 + \cdots + a_n) \quad (5)$$

$$GM = \sqrt[n]{a_1 \cdot a_2 \cdot \cdots \cdot a_n} \quad (6)$$

The most significant individual harmonics (3rd and 5th) are illustrated in Figure 7 and Figure 8. As evident from the figures, the directions of the currents are different.

In this case, mean EV was added to the grid at the residential busbar at different phases one at a time. The resulting voltages and currents are shown in Table 13, voltage THD and PF in Table 14and power values in Table 15.

The EV load causes voltage drop in phase where it is connected more than 5% in each case. Voltage THD was observed to increase up to 0.5% in two cases, but remained at same level when EV was connected to phase C. Power factor changed slightly due to additional EV load and slightly improved in the phase where it was connected.

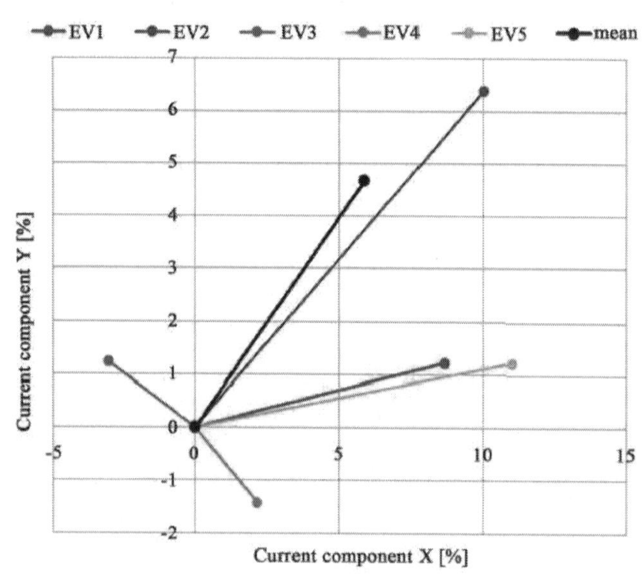

Figure 7. Third current harmonics of EVs.

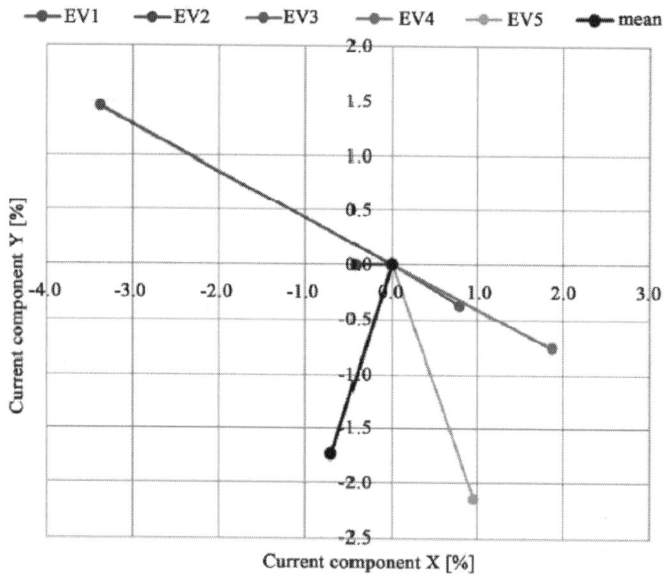

Figure 8. Fifth current harmonics of EVs.

Table 13. Modelled voltage [V] and current [A] values for second case.

Phasing	U_a	U_b	U_c	I_a	I_b	I_c	I_n
EV in A	214	235	235	12.8	1.3	1.3	11.6
EV in B	235	213	236	1.0	13.0	1.3	12.5
EV in C	235	234	215	1.0	1.3	12.8	11.3

Table 14. Modelled voltage THD [%] and PF values for second case.

Phasing	THD_a	THD_b	THD_c	PF_a	PF_b	PF_c
EV in A	3.4	2.1	3.1	0.99	0.92	0.82
EV in B	2.8	2.6	3.0	1.00	1.00	0.85
EV in C	2.9	2.2	3.2	1.00	0.91	1.00

Table 15. Modelled active [kW] and reactive power [kvar] values for second case.

Phasing	P_a	P_b	P_c	Q_a	Q_b	Q_c
EV in A	2.86	0.28	0.24	0.32	−0.11	−0.17
EV in B	0.23	2.91	0.25	−0.02	0.21	−0.16
EV in C	0.23	0.27	2.87	−0.01	−0.13	0.17

C) Third Case—PV and EV

In this case, a PV and EV where both installed at the residential load's busbar. To investigate differences in harmonic cancellation, the PV and EV were installed both on the same phases as well as separate phases.

Voltages and currents are presented in Table 16. It was observed that in arrangements where PV and EV are in same phase voltage drops near to undesirable levels in the phase where the EV is connected while increasing over the 10% limit in the phase where the EV is connected. Also in those cases neutral currents rise much higher than phase currents. The best practice is to install both in same phase. Voltages over the 10% limit and high neutral currents are marked as bold inTable 16.

Voltage THD and PF values are presented in Table 17. In the worst case (EV in A and PV in C), voltage THD decreases in one phase up to 0.6%, but placing EV and PV both on the same phase can lead to slightly decreased voltage THD in some situations (EV in C and PV in C).

Placing the devices onto the same phase leads to slight growth of reactive power in the same phase as evident in Table 18. Active power is compensated due to opposite direction of power flow of the EV and PV.

5. DISCUSSION

While the discreet disturbances of harmonic distortion may not cause immediate and easily-observed impacts, it can cause some equipment to malfunction, and result in additional power losses in both customer and network equipment [28] . As harmonic levels change considerably from one week to another, it is very difficult to assess the long-term evolution of harmonic levels only from measurements carried out over a short period [1]. This paper clearly concludes that power quality problems may occur when dispersed generation and powerful non- linear load utilization is not sufficiently considered. Moreover, modern household devices tend to contain more power electronic circuits and therefore an increase of harmonic currents can also be expected even without the PVs or EVs.

Harmonic current angles of small generators such as PVs or powerful nonlinear loads such as EVs are seldom considered. For determining the impact of current harmonics on the network and the voltage harmonics, the actual harmonic current values would have to be used. One aim of this paper is to draw attention to this topic which could lead to advances in modelling nonlinear generations and loads with different topologies. To help mitigate harmonic distortion problems, models with appropriate harmonic current amplitudes and phase angles could be used to select most suitable devices.

This study only examines one household and one PV or EV at time. The described effects may escalate when a larger number of devices are considered. Special attention is needed in situations where devices have similar harmonic patterns and the harmonic cancellation effect is minimal and grid is weak. General values for further modelling of PVs and EVs were presented in this paper, but additional measurements should be performed to obtain unified values for modelling PV generators and EV loads more accurately. It would be necessary to have measurement data extending over entire years in order to acquire results independent of any disturbance. Furthermore, flicker and voltage level issues should be accounted for as they may have a significant influence in real applications.

Table 16. Modelled voltage [V] and current [A] values for third case.

Phasing	U_a	U_b	U_c	I_a	I_b	I_c	I_n
EV in A and PV in A	231	228	232	2.2	1.3	1.3	3.3
EV in A and PV in B	212	**250**	229	12.9	10.7	1.3	**18.3**
EV in A and PV in C	207	234	**250**	13.1	1.3	10.8	**22.1**
EV in B and PV in A	**250**	206	235	10.9	13.3	1.3	**22.6**
EV in B and PV in B	232	230	229	1.0	1.8	1.3	2.3
EV in B and PV in C	229	211	**251**	1.0	13.1	10.8	**19.5**
EV in C and PV in A	**250**	228	212	10.9	1.3	12.9	**18.8**
EV in C and PV in B	234	**249**	208	1.0	10.8	13.1	**21.4**
EV in C and PV in C	228	231	231	1.0	1.3	1.6	2.5

Table 17. Modelled voltage THD [%] and PF values for third case.

Phasing	THD_a	THD_b	THD_c	PF_a	PF_b	PF_c
EV in A and PV in A	3.2	2.2	3.1	−0.12	0.92	0.84
EV in A and PV in B	3.4	2.1	3.2	0.99	−1.00	0.82
EV in A and PV in C	3.5	2.1	2.9	1.00	0.93	−1.00
EV in B and PV in A	2.6	2.6	3.0	−1.00	1.00	0.86
EV in B and PV in B	2.9	2.4	3.1	1.00	−0.04	0.84
EV in B and PV in C	2.8	2.6	2.8	1.00	1.00	−1.00
EV in C and PV in A	2.6	2.3	3.2	−1.00	0.90	1.00
EV in C and PV in B	2.9	2.2	3.3	1.00	−1.00	1.00
EV in C and PV in C	2.9	2.2	2.9	1.00	0.92	−0.14

Table 18. Modelled active [kW] and reactive power [kvar] values for third case.

Phasing	P_a	P_b	P_c	Q_a	Q_b	Q_c
EV in A and PV in A	−0.06	0.28	0.25	0.51	−0.12	−0.16
EV in A and PV in B	2.86	−2.53	0.25	0.37	0.14	−0.17
EV in A and PV in C	2.91	0.28	−2.58	0.29	−0.11	0.09
EV in B and PV in A	−2.59	2.96	0.25	0.23	0.18	−0.15
EV in B and PV in B	0.23	−0.02	0.25	−0.01	0.42	−0.16
EV in B and PV in C	0.23	2.92	−2.56	−0.02	0.27	0.1
EV in C and PV in A	−2.57	0.28	2.88	0.24	−0.13	0.22
EV in C and PV in B	0.23	−2.55	2.93	0	0.13	0.14
EV in C and PV in C	0.23	0.28	−0.05	−0.01	−0.12	0.37

6. CONCLUSIONS

Firstly, it can be concluded that current harmonic distortion of the PV's output is notably correlated with current and distortion decreases when the PV is operating at a higher loading level. The same conclusion can be made by observing the PF value of the PV which approaches unity with increasing current. Whereas when PF varied considerably, cos(φ) remained near unity throughout the measurement period, which means that main order reactive power is compensated more efficiently than higher order reactive power. EVs showed higher current distortions than PV. Three EVs out of five had current distortion around 10%...12% and two had substantially lower values, around 3%...4%.

Voltage distortion may rise when PVs and EVs are connected to a residential load. The rise is much dependant on grid strength and worse situation may occur in cases where less powerful transformers and thinner cables are utilized. The present study showed that most reasonable option is to connect both PV and EV on same phase. Installing the EVs and PVs separately or installing both on separate phases, results in notable increase in voltage distortion.

Even more severe changes occurred with voltage amplitudes and neutral currents. In cases where PV and EV both were installed, voltage in the phases where PV was connected rose over the 10% limit. Also the neutral currents rose nearly two times higher than phase currents where PV and EV where connected in separate phases. It concludes that modelling with different transformer powers and cable lengths and thicknesses should be conducted before installing nonlinear devices.

REFERENCES

1 Berthet, L., Eyrolles, P., Gauthier, J. and Sabeg, S. (2007) Harmonic Level Measurements on French Low-Voltage Networks. 19th International Conference on Electricity Distribution, Vienna, 21-24 May 2007.

2 Rawa, M.J.H., Thomas, D.W.P. and Sumner, M. (2011) Simulation of Non-Linear Loads for Harmonic Studies. 11th International Conference on Electrical Power Quality and Utilisation (EPQU), Lisbon, 17-19 October 2011, 1-6.

3 Niitsoo, J. and Palu, I. (2011) Distorted Load Impacts on Distribution Grid. Proceedings of the 12th International Scientific Conference Electric Power Engineering 2011, Kouty nad Desnou.

4 Desmet, J., Debruyne, C., Vanalme, J. and Vandevelde, L. (2010) Power Injection by Distributed Generation and the Influence of Harmonic Load Conditions. IEEE Power and Energy Society General Meeting, Minneapolis, 25-29 July 2010, 1-6.

5 Jayasekara, N. and Wolfs, P. (2010) Analysis of Power Quality Impact of High Penetration PV in Residential Feeders. 20th

Australasian Universities Power Engineering Conference (AUPEC), Christchurch.

6 Renders, B., Degroote, L., Driesen, J. and Vandevelde, L. (2007) Profits of Power-Quality Improvement by Residential Distributed Generation. 42nd International Universities Power Engineering Conference, Brighton, 4-6 September 2007, 377-381.

7 Arghandeh, R., Onen, A., Jung, J. and Broadwater, R. (2013) Harmonic Interactions of Multiple Distributed Energy Resources in Power Distribution Networks. Electric Power Systems Research, 105, 124-133.

8 Aiello, M., Cataliotti, A., Favuzza, S. and Graditi, G. (2006) Theoretical and Experimental Comparison of Total Harmonic Distortion Factors for the Evaluation of Harmonic and Interharmonic Pollution of Grid-Connected Photovoltaic Systems. IEEE Transactions on Power Delivery, 21, 1390-1397.

9 Shao, S.P.M. (2011) Demand Response as a Load Shaping Tool in an Intelligent Grid with Electric Vehicles. IEEE Transactions on Smart Grid, 2, 624-631.

10 Kütt, L., Saarijärvi, E., Lehtonen, M. and Niitsoo, J. (2013) A Review of the Harmonic and Unbalance Effects in Electrical Distribution Networks Due to EV Charging. 13th International Conference on Environment and Electrical Engineering (EEEIC), Wroclaw.

11 Yong, J., Chen, L. and Chen, S. (2010) Modeling of Home Appliances for Power Distribution System Harmonic Analysis. IEEE Transactions on Power Delivery, 25, 3147-3155.

12 Jensen, M.M. and Hansen, H. (2014) Triplen Harmonic Voltage Distortion in the Low-Voltage Network. NORDAC 2014, Stockholm.

13 Block, P., Salamanca, H., Teixeira, M., Dahlke, D., Shiono, O. and Dodadon, A. (2014) Power Quality Analyses of a Large Scale Photovoltaic System. The fifth International Renewable Energy Congress IREC 2014, Hammamet, 25-27 March 2014, 1-6.

14 Kontogiannis, K., Vokas, G., Nanou, S. and Papathanassiou, S. (2013) Power Quality Field Measurements on PV Inverters. International Journal of Advanced Research in Electrical Electronics and Instrumentation Engineering, 2, 5301-5314.

15 Ortega, M., Hernández, J. and García, O. (2013) Measurement and Assessment of Power Quality Characteristics for Photovoltaic Systems: Harmonics, Flicker, Unbalance, and Slow Voltage Variations. Electric Power Systems Research, 96, 23-35.

16 Anwari, M., Hamid, M. and Taufik (2008) Power Quality Behavior of Single Phase Fed Adjustable Speed Drive Supplied from Grid of PV Generation. The 2nd IEEE International Conference on Power and Energy (PECon 08), Bahru, 1-3 December 2008, 1098-1102.

17 van Lumig, M., Bhattacharyya, S., Gobben, J. and Kling, W. (2010) Power Quality Measurements Near DER and Disturbing Loads. International Conference on Renewable Energies and Power Quality (ICREPQ'11), Las Palmas de Gran Ganaria, 13-15 April 2010.

18 Wang, Y.J., O'Connell, R. and Brownfield, G. (2001) Modeling and Prediction of Distribution System Voltage Distortion Caused by Nonlinear Residential Loads. IEEE Transactions on Power Delivery, 16, 744-751.

19 Berisha, S., Karady, G., Ahmad, R., Hobbs, R. and Karner, D. (1996) Current Harmonics Generated by Electric Vehicle Battery Chargers. Proceedings of the 1996 International Conference on Power Electronics, Drives and Energy Systems for Industrial Growth, New Delhi, 8-11 January 1996, 584-589.

20 Zamri, M., Wanik, C., Fadzil, M., Siam, M., Ayob, A., Yanto, S., Azah, M., Azit, A., Sulaiman, S., Azrin, M., Ali, M., Hussein, Z.F., Kamil, A. and Hussin, M. (2013) Harmonic Measurement and Analysis during Electric Vehicle Charging. Engineering, 5, 215-220.

21 Melo, N., Mira, F., de Almeida, A. and Delgado, J. (2011) Integration of PEV in Portuguese Distribution Grid: Analysis of Harmonic Current Emissions in Charging Points. 11th International Conference on Electrical Power Quality and Utilisation (EPQU), Lisbon, 17-19 October 2011, 1-6.

22 Wenge, C., Stotzer, M., Winkler, T. and Komarnicki, P. (2011) Power Quality Measurements of Electric Vehicles in the Low Voltage Power Grid. 11th International Conference on Electrical Power Quality and Utilisation (EPQU) Lisbon, 17-19 October 2011, 1-5.

23 Fekete, K., Klaic, Z. and Majdandzic, L. (2012) Expansion of the Residential Photovoltaic Systems and Its Harmonic Impact on the Distribution Grid. Renewable Energy, 43, 140-148.

24 McBee, K. and Simoes, M. (2012) Evaluating the long-Term Impact of a Continuously Increasing Harmonic Load Demand on Feeder Level Voltage Distortion. 2012 IEEE Industry Applications Society Annual Meeting (IAS), Las Vegas, 7-11 October 2012, 1-8.

25 Au, M.T., Navamany, J. and Yeoh, E.C. (2010) Impact of Medium Power Inverter-Based Residential Harmonic Loads and PFCs on Low Voltage Lines. 2010 IEEE International Conference on Power and Energy (PECon), Kuala Lumpur, 29 November-1 December 2010, 695-699.

26 EN 50160:2010 (2010) Voltage Characteristics in Public Distribution Systems. European Standard.

27 Niitsoo, J., Kilter, J., Palu I., Taklaja, P. and Kütt, L. (2013) Harmonic Levels of Domestic and Electrical Vehicle Loads in Residential Distribution Networks. Proceedings of AFRICON 2013, Mauritius, 9-12 September 2013, 1-5.

28 Niitsoo, J., Palu, I., Kilter, J., Taklaja, P. and Vaimann, T. (2013) Residential Load Harmonics in Distribution Grid. Proceedings of 3rd International Conference on Electric Power and Energy Conversion Systems (EPECS 2013), Istanbul, 2-4 October 2013, 1-6.

CHAPTER 11

Fault Diagnosis in Distribution Networks with Distributed Generation

Amany M. El-Zonkoly

Department of Electrical & Control Engineering, Collage of Engineering & Technology, Arab Academy for Science & Technology, Miami, Egypt.

ABSTRACT

The penetration of distributed generation (DG) in distribution power system would affect the traditional fault current level and characteristics. Consequently, the traditional protection arrangements developed in distribution utilities are difficult in coordination. Also, the reclosing scheme would be affected. With the rapid developments in distribution system automation and communication technology, the protection coordination and reclosing scheme based on information exchange for distribution power system can be realized flexibly. This paper proposes a multi-agent based scheme for fault diagnosis in power distribution networks with distributed generators. The relay agents are located such that the distribution network is divided into several sections. The relay agents measure the bus currents at which they are located such that it can detect and classify the fault, and determine the fault location. The proposed technique uses the entropy of wavelet coefficients of the measured bus currents. The performance of the proposed protection scheme is tested through simulation of two systems. The first system is a benchmark medium voltage (MV) distribution system and the second system is practical 66 kV system of the city of Alexandria.

KEYWORDS

Cooperative Systems, Relay Agent, Fault Diagnosis, Distributed Generation, Wavelet Transform, Entropy Calculation

1. INTRODUCTION

The traditional power systems, which are based on large fossil fuel fired power generation plants, long distance transmission lines and hierarchical control centers, are changing. A large number of distributed generation units including renewable energy sources such as wind turbines, PV generators, fuel cells together with Combined Heat and Power (CHP) plants, are being integrated into power systems at distribution level. The penetration of DGs changes the traditional distribution power system short circuit power, fault current level and the characteristics of the fault current, such as amplitude, direction and distribution [1].

Most distribution network protection schemes are initially designed without DGs. The nature of distribution network can be radial or meshed. Traditionally, radial networks are protected using coordinated overcurrent relays whereas meshed networks are protected using directional overcurrent relays [2].

To deal with the problem of protection of distributed networks with the penetration of DGs, distribution utilities impose interconnected regulations. These regulations are often based on IEEE Std. 1547, 2003 [4] and recommend the tripping of DGs even for remote faults in order to maintain the protection coordination during fault. According to [4], immediate tripping of DGs is recommended also if a power island is created. That is why protection schemes, which are capable of immediate identifying and isolating faults after their occurrence, are required. These schemes can ease the requirement for the disconnection of DGs to ensure protection coordination and enable intentional power islands.

Several protection schemes have been proposed in literature [1-8] in order to address the shortcomings in current DG interconnection practices and protection problems associated with DGs.

This paper proposes a multi-agent based protection scheme to classify and locate the fault in a distribution network with DG. The proposed technique aims to locate and isolate a faulty section in a distribution system with DGs. It is based on entropy calculation of wavelet coefficients of the three phase current signals. This method uses only current signals measured by relay agents at the boundaries of the network sections to identify the type of fault if it is a three line to ground (3LG), single line to ground (LG), double line to ground (DLG) or a line to line (LL) fault. It also determines the phases included in fault and the bus or line at which the fault occurred. The performance of the proposed algorithm is investigated through the simulation of a benchmark MV distribution system and part of the 66 kV network if Alexandria. The results proved the effectiveness of the proposed protection scheme under different conditions of fault type, fault location and fault resistance.

2. ANALYSIS OF THREE PHASE POWER SYSTEM TRANSIENTS

2.1. Modal Transformation

In three-phase systems, many different fault types depending on the phase involved or the involvement of ground can occur. In order to diagnose such types of faults, currents and/or voltages of all three-phase quantities must be analyzed. However, the amount of processing can be reduced by transforming three-phase quantities into modal components.

The modal transformation resolves three-phase signals in a coupled network into three uncoupled modal components, namely, 1) the ground mode; 2) aerial mode-1; and 3) aerial mode-2 components. For nontransposed multiphase systems, an eigenvector-based frequency-dependent transformation matrix is required to convert the quantities from phase domain to modal domain. For balanced and ideally transposed lines, a frequency-independent, real transformation matrix, such as Clarke transformation, can be used. Although practical distribution systems do not satisfy the aforementioned conditions, a frequency-independent real transformation matrix can be used to obtain somewhat decoupled signals that can be advantageous in transient-based fault location.

The relationship between the Clarke components and the phase components is given by

$$\begin{pmatrix} I_0 \\ I_\alpha \\ I_\beta \end{pmatrix} = \frac{1}{3}\begin{pmatrix} 1 & 1 & 1 \\ 2 & -1 & -1 \\ 0 & \sqrt{3} & -\sqrt{3} \end{pmatrix}\begin{pmatrix} I_a \\ I_b \\ I_c \end{pmatrix} \tag{1}$$

where, I_a, I_b and I_c are the phase currents and I_0, I_α and I_β are the respective Clarke components. Transients in the phase currents are well reflected in the Clarke components.

2.2. Wavelet Transformation and Entropy Calculation

Lots of fault information is included in the transient components. So it can be used to identify the fault or abnormity of equipments or power system. It can also be used to deal with the fault and analyze its reason. This way the reliability of the power system will be considerably improved.

Transient signals have some characteristics such as high frequency and instant break. Wavelet transform is capable of revealing aspects of data that other signal analysis techniques miss and it satisfies the analysis need of electric transient signals. Usually, wavelet transform of transient signal is expressed by multi-revolution decomposition fast algorithm which utilizes the orthogonal wavelet bases to decompose the signal to components under different scales. It is equal to recursively filtering the signal with a high-pass and low-pass filter pair. The approximations are the high-scale, low-frequency components of the signal produced by filtering the signal by a low-pass filter. The details are the low-scale, high-frequency components of the signal produced by filtering the signal by a high-pass filter. The band width of these two filters is equal. After each level of decomposition, the sampling frequency is reduced by half. Then recursively decompose the low-pass filter outputs (approximations) to produce the components of the next stage [9,10].

Given a discrete signal x(n), being fast transformed at instant k and scale j, it has a high-frequency component coefficient $D_j(k)$ and a low-frequency component coefficient $A_j(k)$. The frequency band of the information contained in signal components $D_j(k)$ and $A_j(k)$, obtained by reconstruction are as follows [11].

$$\begin{cases} D_j(k):\left[2^{-(j+1)} f_s, 2^{-j} f_s\right] \\ A_j(k):\left[0, 2^{-(j+1)} f_s\right] \end{cases} (j=1,2,\cdots,m) \tag{2}$$

where, f_s is the sampling frequency.
The original signal sequence x(n) can be represented by the sum of all components as follows [11].

$$\begin{aligned} x(n) &= D_1(n)+A_1(n)=D_1(n)+D_2(n)+A_2(n) \\ &= \sum_{j=1}^{J} D_j(n)+A_J(n) \end{aligned} \tag{3}$$

Various wavelet entropy measures were defined in [9]. In this paper, the nonnormalized Shannon entropy will be used. The definition of nonnormalized Shannon entropy is as follows [11].

$$E_j = -\sum_k E_{jk} \log E_{jk} \tag{4}$$

Where E_{jk} is the wavelet energy spectrum at scale j and instant k and it is defined as follows.

$$E_{jk} = \left|D_j(k)\right|^2 \tag{5}$$

3. PROPOSED AGENT-BASED FAULT DIAGNOSIS

3.1. Protection Arrangement Based on Relay Agents

Protective relays detect fault occurrence in a power system and isolate that part of the power system to prevent fault from affecting the whole power system. Traditional protective schemes generally used dual systems of primary and backup protection relays for high sensitive and reliable protection of the system. The primary protection relays are usually current differential relaying that has a high accuracy of fault detection. The backup protection relays are usually distance relaying that work with local power system information only [12].

With the introduction of distributed generation and deregulation, the power system impedance and fault currents through protective devices would change. The protective devices are therefore difficult to be coordinated [13].

The distribution power system automation techniques have been widely adopted and the infrastructure of communication has been developed. The protection schemes based on microprocessors with communication capabilities are utilized, so that the status of the relays and breakers can be obtained from the distribution power system supervisory control and data acquisition system, which can serve as an information exchange platform. Based on the platform, the protection coordination and adaptation can be dealt with flexibly [1].

The concept of a cooperative protection system with an agent model (relay agent) was first proposed by Y. Tomita, et al. back in 1998 [12].

The application of this concept has been proposed in [14] and [13] to build an intelligent, adaptive protection system.

The relay agents allow the cooperation between numbers of equipments. The cooperation of relay agents enables the flexible utilization of all hardware resources by any equipment, so that various adaptive protective functions are realized.

The fundamental concepts of relay agents as proposed by Y. Tomita, et al. [12] are as follows:

1) Equipment Agent This type of agent manages data and functions proper to each pieces of equipment. Network topology data of a power system are expressed as link data between equipment agents. This enables distribution of power system information.

2) Mobile Agent This type of agent moves between equipment agents or protection system equipment to utilize data and functions distributed in equipment agents.

3) Protector Agent This type of agent dispatches mobile agents to detect a fault or isolate a fault-detected zone. These agents act as supervisors for detection and isolation.

4) Reorganizer Agent This type of agent reorganizes the agent space according to changes in power system conditions. The cooperative protection system with relay agents must have an agent processing facility, which allows the agent to transfer to the different hardware through communication networks or to communicate with other agent in the different hardware.

In this paper, the distribution network is divided into a number of network segments as shown in**Figure 1** for fault isolation purpose. Each network segment can be isolated from the rest of the system for a fault inside it by opening the circuit breakers (CBs) at its interconnection points to the system. The three phase bus currents are measured and fed to the relay agents which are placed at the interconnection points of different network segments. The relay agents exchange data between each other through a telecommunication network. A simple algorithm based on the information collected by different relay agents is proposed to classify and then locate the fault such that the faulted area can be correctly isolated.

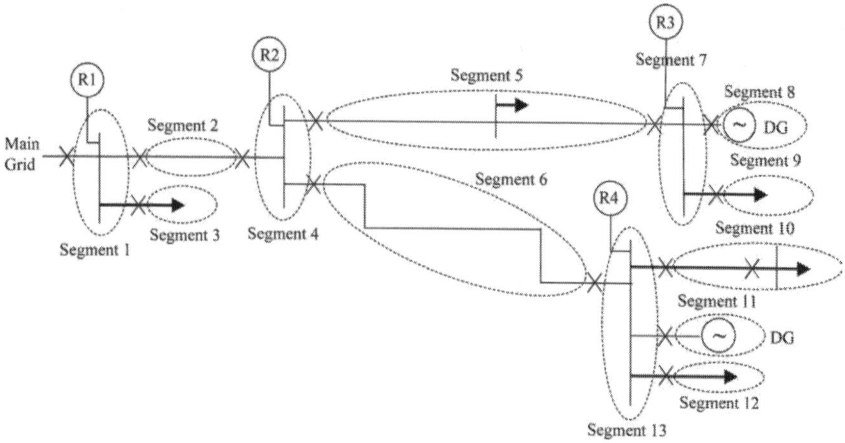

Figure 1. Structure of proposed protection scheme.

3.2. Fault Classification Using Current Transients

An analysis of all possible types of fault in three phase system, i.e. LG faults (AG, BG, CG), LL faults (AB, BC, CA), DLG faults (ABG, BCG, CAG) and 3LG faults (ABCG), is carried out. In this paper, the proposed algorithm determines the type of fault first, then the phases included in fault and finally it determines the fault location.

The sum of absolute entropies of wavelet coefficients of the Clarke component I_0 is used to determine the type of fault if it is 3LG, LG, DLG, or LL fault. Next, the sum of absolute entropies of wavelet coefficients of the three phase currents (I_a, I_b, and I_c) are used to determine the phases included in fault. Finally, the values of the sum of absolute entropies of wavelet coefficients of Clarke components I_a and I_β are used to locate the fault. The mother wavelet used was 'Symlets' in addition to Shannon entropy.

4. SIMULATION AND RESULTS

4.1. Test System 1

The distribution system shown in **Figure 2** was simulated using the SIMULINK power system blockset. This system has been derived from the CIGRE MV benchmark test distribution system [15]. The system used in the study has two DGs. DG-1 is a 1500-kVA three-phase induction generator running from a wind turbine and DG-2 is a 200-kVA small-size three-phase synchronous generator running from a hydro turbine. The generators were simulated using detailed machine models. Distribution lines–lengths of which range from 0.24 km to 4.89 km–were modeled using transmission line p sections. The loads, which include highly unbalanced three-phase loads as well as single-phase loads, were modeled using series RLC loads. Each relay agent takes the measurements from the corresponding busbar located near the relay agent. Locations of the relay agents were determined based on the network configuration, interconnection to the grid, and locations of the DGs. It was assumed that the relay samples current signals at a frequency of 10 kHz.

4.1.1. Radial Network

In the proposed CIGRE MV benchmark network, switches S2 and S3 were kept open to simulate a radial network.

For a fault at bus 3, the sum of absolute entropy of wavelet coefficients of I_0 monitored at relay agent R1for different fault types are listed in **Table 1**. The fault was simulated from 10/60 to 15/60 s with fault resistance $R_f = 10$ W.

From the values of sumI_0 shown in **Table 1**, the fault type if it was a 3LG or a LL fault is well defined. In case of LG or DLG faults, sumI_0 has nearly the same range of values. In that case sum$I_a$and sumI_β are used to discriminate between a LG fault and a DLG fault. The values of sumI_a and sumI_β for currents monitored at relay agent R1 are given in **Table 2**.

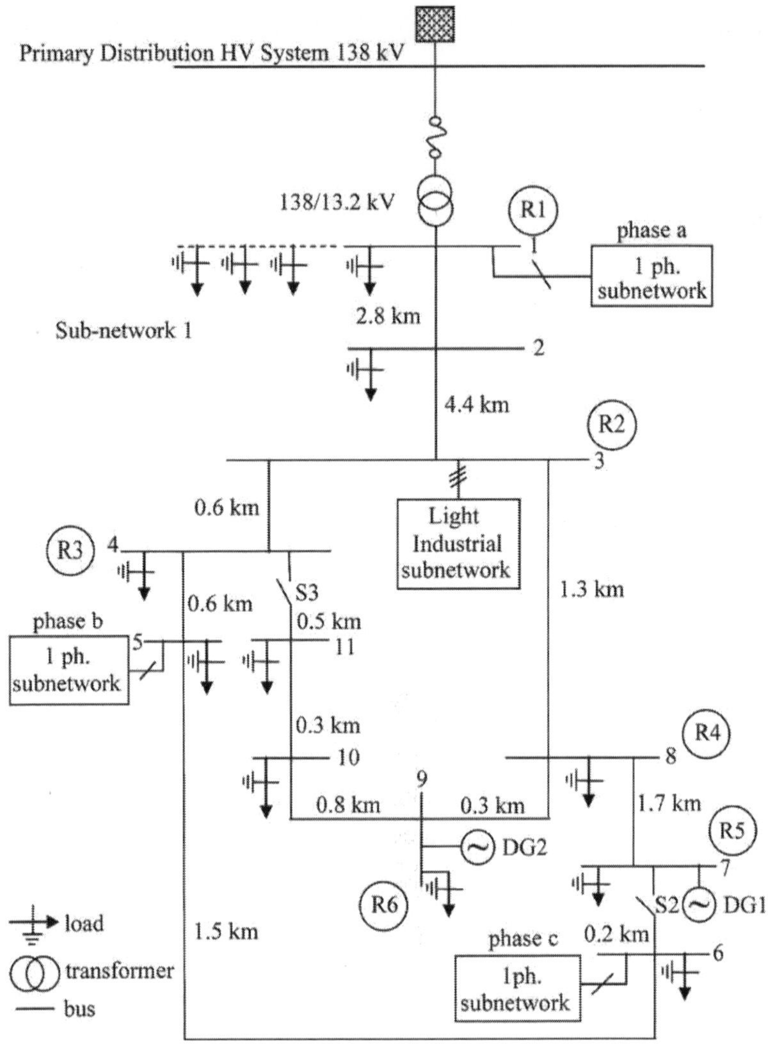

Figure 2. MV test distribution system.

As shown in **Table 2**, the maximum value of sumI_α and sumI_β in case of LG fault is less than 30,000 and in case of DLG fault it is more than 50,000. This way these two types of fault are discriminated from each other.

The next step in the proposed algorithm after determining the fault type is to determine the phases included in fault. To do so the sum of entropies of wavelet coefficients of the three phase currents are used. For the same fault at bus 3, **Table 3** lists the values of $sumI_a$, $sumI_b$ and $sumI_c$ for currents monitored at relay agent R1 in case of LG, LL and DLG faults.

Table 1. The sum of absolute entropy of wavelet coefficients of I_0monitored at R1 for a fault at bus 3.

Fault Type	ABCG	AG	BG	CG	AB	BC	CA	ABG	BCG	CAG
$sumI_0$	5.83	56.5	51.4	59.6	0.001	0.001	0.001	58.2	61.3	65

Table 2. The values of $sumI_a$ and $sumI_\beta$ of currents monitored at R1 for a fault at bus 3.

Fault Type	AG	BG	CG	ABG	BCG	CAG
$sumI_\alpha$	20,255	3,068	2,950	60,200	2,917	51,700
$sumI_\beta$	260	8,398	8,854	9,900	50,700	13,000

Table 3. The values of $sumI_a$, $sumI_b$ and $sumI_c$ of currents monitored at R1 for a fault at bus 3.

Fault Type	AG	BG	CG	AB	BC	CA	ABG	BCG	CAG
$sumI_a$	26,510	1,615	1,214	57,375	270	49,967	66,279	1,525	49,327
$sumI_b$	381	16.179	492	32,565	33,779	200	30,656	40,536	389
$sumI_c$	599	1,169	20,045	338	42,136	35,962	587	40,375	44,302

As shown from the results in **Table 3**, in case of LG fault, the faulted phase is that having the maximum sumI. For example, for an AG fault, $sumI_a$ is greater than $sumI_b$ and $sumI_c$. In case of LL fault, the phase of minimum sumI is not included. For example, for an BC fault, $sumI_a$ is less than both $sumI_b$ and $sumI_c$. Finally, in case of DLG fault, also the phase with minimum sum is not included in fault. For example, for an CAG fault, $sumI_b$ is less than $sumI_a$ and $sumI_c$.

The final step of the algorithm is to determine the fault location. To determine the fault location in case of 3LG and DLG faults the values of

$sumI_0$ of currents monitored at different relay agents are used. For example, in case of 3LG at buses 4, 5 or 6, the value of $sumI_0$ for currents monitored at relay agent R3 is used to determine the faulted bus. Where, in that case $sumI_0 > 10$ and if the fault at any other bus $sumI_0 < 10$. To determine the exact fault location the value of $sumI_0$ is used. For fault at bus 4, $sumI_0 = 55.6$. For fault at bus 5, $sumI_0 = 35.7$. For fault at bus 6, $sumI_0 = 10.4$. **Table 4** lists the values of $sumI_0$ at R3 for 3LG fault at all buses.

For a DLG fault at buses 9, 10 or 11, $sumI_0$ for currents monitored at relay agent R6 is greater than 500. In the same time, for the same type of fault at any other bus, $sumI_0$ at R6 is less than 1. To determine the exact location of the fault, both the values of $sumI_0$ at relay agent R6 and $sumI_0$ at relay agent R5 are used. **Table 5** lists the values of $sumI_0$ at R5 and at R6 for DLG fault at buses 9, 10 and 11.

Table 4. The values of $sumI_0$ of currents monitored at R3 for an 3LG fault at different buses.

Fault Location	$sumI_0$ – R3
1	0.0054
2	0.0046
3	0.0052
4	55.665
5	35.76
6	10.39
7	0.0047
8	0.0071
9	0.0052
10	0.0047
11	0.0048

In case of LL fault, to determine the faulted bus, the values of sumI_α and sumI_β for currents monitored at different relay agents are used. For example, for an AB fault, the fault location is determined by monitoring sumI_α at relay agents R1, R3 and R4. **Table 6** lists the values of sum$I_\alpha$at R1, R3 and R4 for an AB fault at different buses.

In case of LG fault, sumI_α and sumI_0, monitored at different relay agents, are used to determine the fault location. For example, for an AG fault, the values of sumI_0 at R1, R3, R4 and at R6 are listed in **Table 7**.

As shown in **Table 7**, for an AG faults at buses 1 and 2, sumI_0 at R1 > 300. For faults at buses 4, 5 and 6, sumI_0 at R3 > 700. For faults at buses 9, 10 and 11, sumI_0 at R6 > 1000. For faults at buses 3, 7 and 8, the previous values doesn't apply but sumI_0 at R4 for a fault at bus 3 is > 400. In the same way, LG faults including the two other phases could be located.

Also, for faults at different branches connecting two buses, the entropy sum of wavelet coefficients of three phase currents and their Clarke components can identify the type of fault, phases included in fault and fault location. For example, for a 3LG fault at line 3-8, the values of sumI_α and sumI_0monitored at relay agent R1 comes between their values for the same fault at bus 3 and at bus 8. Although sumI_α and sumI_0 monitored at R1 for 3LG fault at bus 4 or 5 comes also between those values at bus 3 and at bus 8, the value of sumI_0 at R3 > 10, which is not the case for fault at bus 3, bus 8 or line 3-8. **Table 8** lists the values of sumI_α and sumI_0 as monitored at R1 and sumI_0 as monitored at R3 for 3LG fault at buses 3, 4, 5, 8 and line 3-8.

For all cases of fault at different locations, the proposed protection algorithm correctly recognized the type of fault and the faulty segments even with high fault resistance (R_f = 250 W).

4.1.2. Meshed Network

In order to test the proposed protection algorithm in meshed networks, the switches S2 and S3 in the MV network shown in **Figure 2** were closed. The same faults simulated in case of the radial system were applied to the meshed network too.

Table 5. The values of sumI_0 of currents monitored at R5 and R6 for DLG faults at buses 9, 10 and 11.

Fault Type		ABG	BCG	CAG
Bus 9	R5	191.22	581.83	177.36
	R6	1329.6	1040.4	2309.5
Bus 10	R5	67.403	269.75	77.469
	R6	767.13	670.17	1274.6
Bus 11	R5	38.821	201.29	51.447
	R6	629.5	566.3	1037.9

Table 6. The values of sumI_a of currents monitored at R1, R3 and R4 for an AB fault at different buses.

Fault Location	sumI_α – R1	sumI_α – R3	sumI_α – R4
1	6E+06	0.0302	199.31
2	326699	0.0317	181.31
3	57378	0.0338	169.65
4	48217	53945	179.13
5	41035	45216	186.75
6	28668	30448	204.97
7	26854	0.0292	26823
8	40323	0.0368	40285
9	37279	0.0336	37242
10	30586	0.03	30552
11	28522	0.0305	28489

Table 7. The values of sumI_0 of currents monitored at R1, R3, R4 and R6 for an AG fault at different buses.

Fault Location	sumI_0 – R1	sumI_0 – R3	sumI_0 – R4	sumI_0 – R6
1	1889	0.0044	346	0.0001
2	336	0.0043	311	0.0001
3	56	0.0071	451	0.0026
4	60	2138	318	0.0006
5	61	1583	222	0.0011
6	56	798	78	0.001
7	43	0.0047	43	0.0016
8	60	0.0051	60	0.0051
9	58	0.0046	58	2713
10	54	0.0053	54	1766
11	52	0.0055	52	1519

In the same way as in case of radial network, the entropy sum of wavelet coefficients of three phase currents are used to select the phases included in fault. On the other hand, the entropy sum of wavelet coefficients of Clarke components are used to determine the type of fault and its location. Using the entropy calculation through the proposed algorithm made the algorithm very sufficient and correctly succeeded in classifying and locating the fault during all conditions of simulation. That is considered as an improvement over the algorithm proposed in [3], which failed to locate the fault in several cases. Tables 9-11 list some results monitored in case of meshed network.

For a fault at bus 3, the sum of absolute entropy of wavelet coefficients of I_0 monitored at relay agent R1for different fault types are listed in **Table 9**. The fault was simulated from 10/60 to 15/60 s with fault resistance $R_f = 10$ W.

From the values of sumI_0 shown in **Table 9**, the fault type if it was a 3LG or a LL fault is well defined. In case of LG or DLG faults, sumI_0 has nearly the same range of values. In that case sum$I_a$and sumI_β are used to discriminate between a LG fault and a DLG fault. The values of $sumI_a$ and $sumI_\beta$ for currents monitored at relay agent R1 are given in **Table 10**.

Table 8. The values of $sumI_a$ and $sumI_0$ of currents monitored at R1 and R3 for 3LG faults at different buses

Fault Location	$sumI_\alpha$ – R1	$sumI_0$ – R1	$sumI_0$ – R3
Bus 3	78077	5.8	0.0052
Bus 4	65873	5.4	55.7
Bus 5	56272	4.9	35.8
Bus 8	55322	4.8	0.0071
Line 3-8	65360	5.4	0.0045

As shown in **Table 10**, the maximum value of $sumI_a$ and $sumI_\beta$ in case of LG fault is less than 1200 and in case of DLG fault it is more than 1600. This way these two types of fault are discriminated from each other.

The next step in the proposed algorithm after determining the fault type is to determine the phases included in fault. To do so the sum of entropies of wavelet coefficients of the three phase currents are used. For the same fault at bus 3, **Table 11** lists the values of $sumI_a$, $sumI_b$ and $sumI_c$ for currents monitored at relay agent R1 in case of LG, LL and DLG faults.

As shown from the results in **Table 11**, in case of LG fault, the faulted phase is that having the maximum sumI. For example, for an AG fault, $sumI_a$ is greater than $sumI_b$ and $sumI_c$. In case of LL fault, the phase of minimum sumI is not included. For example, for an BC fault, $sumI_a$ is less than both $sumI_b$ and $sumI_c$. Finally, in case of DLG fault, also the phase with minimum sum is not included in fault. For example, for an CAG fault, $sumI_b$ is less than $sumI_a$ and $sumI_c$.

4.2. Test System 2

The test system shown in **Figure 3** was simulated using the SIMULINK power system blockset. This system is a part of the 66 kV network of the city of Alexandria, Egypt.

The system used in the study has four DGs each of them are 1 MVA three-phase synchronous generator driven by diesel engines. The system line and load data are give in the Appendix. Each relay agent takes the measurements from the corresponding busbar located near the relay agent. Locations of the relay agents were determined based on the network configuration and locations of the DGs. It was assumed that the relay samples current signals at a frequency of 20 kHz.

In the proposed 66 kV network the following results were taken. For a fault at bus 5, the sum of absolute entropy of wavelet coefficients of I_0 monitored at relay agent R3 for different fault types are listed in **Table 12**. The fault was simulated from 10/50 to 15/50 s with fault resistance $R_f = 10$ W.

From the values of sumI_0 shown in **Table 12**, the fault type if it was a 3LG or a LL fault is well defined. In case of LG or DLG faults, sumI_0 has nearly the same range of values. In that case sum$I_a$and sumI_β are used to discriminate between a LG fault and a DLG fault. The values of sumI_a and sumI_β for currents monitored at relay agent R3 are given in **Table 13**.

As shown in **Table 13**, the maximum value of sumI_a and sumI_β in case of LG fault is less than 70,000 and in case of DLG fault it is more than 70,000. This way these two types of fault are discriminated from each other.

Table 9. The sum of absolute entropy of wavelet coefficients of I_0monitored at R1 for a fault at bus 3 in meshed network.

Fault Type	ABCG	AG	BG	CG	AB	BC	CA	ABG	BCG	CAG
sumI_0	0.54	21	23	23	0.0011	0.0011	0.0011	22	22	22

Table 10. The values of $sumI_\alpha$ and $sumI_\beta$ of currents monitored at R1 for a fault at bus 3 in meshed network.

Fault Type	AG	BG	CG	ABG	BCG	CAG
$sumI_\alpha$	1084	205	309	1784	80	2773
$sumI_\beta$	242	371	53	429	1936	73

Table 11. The values of $sumI_a$, $sumI_b$ and $sumI_c$ of currents monitored at R1 for a fault at bus 3 in meshed network.

Fault Type	AG	BG	CG	AB	BC	CA	ABG	BCG	CAG
$sumI_a$	1379	213	128	1512	227	2801	2176	2	2540
$sumI_b$	139	523	173	1494	603	163	1317	1063	172
$sumI_c$	337	156	906	339	2228	1004	213	1998	1642

The next step in the proposed algorithm after determining the fault type is to determine the phases included in fault. To do so the sum of entropies of wavelet coefficients of the three phase currents are used. For the same fault at bus 5, **Table 14** lists the values of $sumI_a$, $sumI_b$ and $sumI_c$ for currents monitored at relay agent R3 in case of LG, LL and DLG faults.

As shown from the results in **Table 14**, in case of LG fault, the faulted phase is that having the maximum sumI. For example, for an AG fault, $sumI_a$ is greater than $sumI_b$ and $sumI_c$. In case of LL fault, the phase of minimum sumI is not included. For example, for an BC fault, $sumI_a$ is less than both $sumI_b$ and $sumI_c$. Finally, in case of DLG fault, also the phase with minimum sum is not included in fault. For example, for an CAG fault, $sumI_b$ is less than $sumI_a$ and $sumI_c$.

The final step of the algorithm is to determine the fault location. To determine the fault location in case of 3LG and DLG faults the values of $sumI_0$ of currents monitored at different relay agents are used. For example, in case of 3LG, the value of $sumI_0$ for currents monitored at relay agent R5 is used to determine the faulted bus. **Table 15** lists the values of $sumI_0$ at R5 for 3LG fault at all buses. As shown in **Table 15**, the values of $sumI_0$ at all buses are distinguished from each other except at buses 4, 5, 6 and 8. For the fault at these buses the value of $sumI_0$ at R7 are used to determine the fault location.

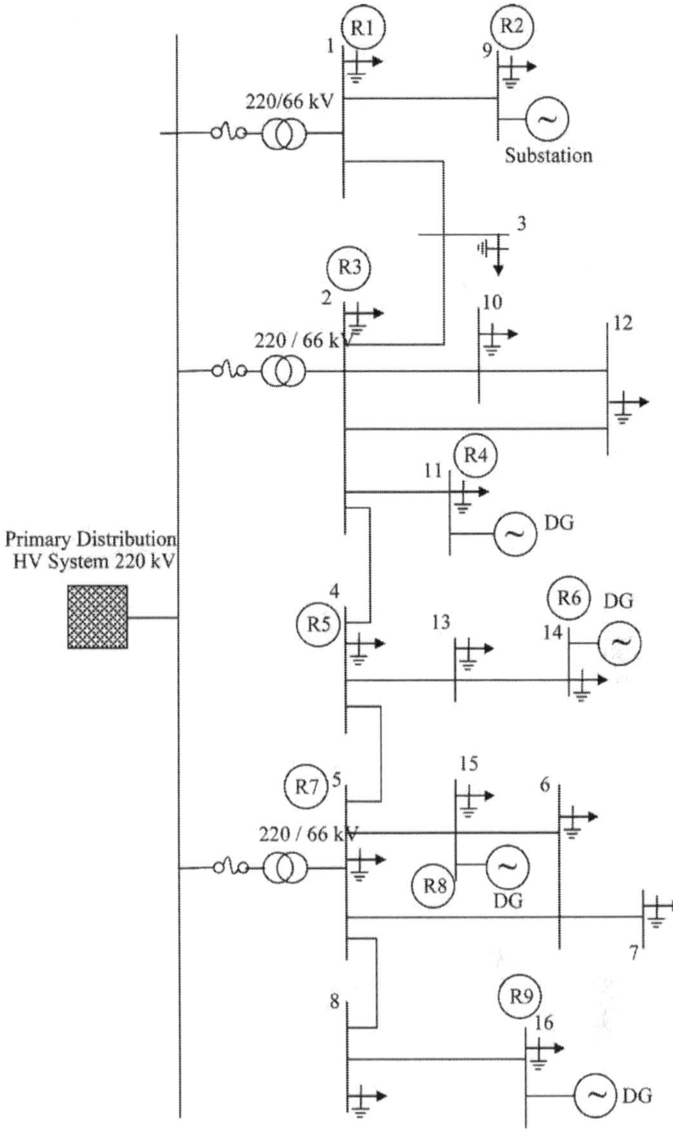

Figure 3. Alexandria 66 kV network.

Table 12. The sum of absolute entropy of wavelet coefficients of I_0monitored at R3 for a fault at bus 5.

Fault Type	ABCG	AG	BG	CG	AB	BC	CA	ABG	BCG	CAG
$sumI_0$	0.0565	13.124	13.248	18.686	5.4e-9	5.7e-9	6.4e-9	11.412	13.032	13.269

Table 13. The values of sumI_α and sumI_β of currents monitored at R2 for a fault at bus5.

Fault Type	AG	BG	CG	ABG	BCG	CAG
sumI_α	68199	60553	62672	70359	63391	75148
sumI_β	59108	67427	64685	68701	76200	63733

Table 14. The values of sumI_a, sumI_b and sumI_c of currents monitored at R3 for a fault at bus 5.

Fault Type	AG	BG	CG	AB	BC	CA	ABG	BCG	CAG
sumI_a	69868	59781	61682	68541.1	59114.5	74495.5	70789	61949	75828
sumI_b	61448	70165	59695	73724.6	68875.7	59144.7	75775	71058	61625
sumI_c	60348	63131	74579	59720.8	81133.1	72128.4	63361	82786	75587

Table 15. The values of sumI_0 of currents monitored at R5 and R7 for an 3LG fault at different buses.

Fault Location	sumI_0 – R5	sumI_0 – R7
1	9.05	5.24
2	7.52	6.11
3	9.87	4.78
4	0.045	6.66
5	0.039	0.009
6	0.038	0.03
7	3.62	7.43
8	0.042	0.013
9	3.9	1.193
10	0.02	0.015
11	0.4	0.161
12	0.57	3.29
13	0.01	0.013
14	0.18	5.13
15	4.01	7.85
16	0.07	0.039

For a DLG fault at any bus, sumI_0 for currents monitored at relay agent R5 is used. **Table 16** lists the values of sumI_0 at R5 at different buses.

In case of LL fault, to determine the faulted bus, thevalues of sumI_a and sumI_β for currents monitored at different relay agents are used. For example, for an AB fault, the fault location is determined by monitoring sumI_a at relay agents R2, R3 and R7. **Table 17** lists the values of sumI_a at R2, R3 and R7 for an AB fault at different buses. For near similar readings at any one of the relay agents the other two agent readings are used to locate the faulted bus.

In case of LG fault, sumI_a and sumI_0, monitored at different relay agents, are used to determine the fault location. For example, for an AG fault, the values of sumI_0 at R3, R4 and at R5 are listed in **Table 18**.

As shown in **Table 18**, if the values of sumI_0 at relay agent R3 is similar for AG fault at more than one bus as in the case of buses 3, 7 and 13, the values of sumI_0 at relay agent R4 is used to identify the faulted bus.

Table 16. The values of sumI_0 of currents monitored at R5 DLG faults at different buses.

Bus	ABG	BCG	CAG
1	24.5	24.8	25.2
2	4.12	5.2	8.23
3	35.1	35.5	36.1
4	26.3	26.5	27.1
5	31.1	31.8	32.4
6	19.2	19.5	19.9
7	11.6	11.8	12.1
8	27.6	27.8	28.1
9	12.4	12.5	13
10	18	18.4	18.8
11	16.1	16.3	16.8
12	20.2	20.8	21.3
13	15	15.4	15.7
14	32.7	32.9	33.3
15	13.2	13.6	13.9
16	21.4	21.7	22

Table 17. The values of sumI_a of currents monitored at R2, R3 and R7 for an AB fault at different buses.

Fault Location	sumI_α – R2	sumI_α – R3	sumI_α – R7
1	59246	18909	4760
2	77414	18968	4743
3	69921	18877	4738
4	73488	19016	4737
5	68541	36716	4733
6	68330	33808	4724
7	66826	30364	4720
8	68773	35129	4729
9	59118	18849	16469
10	70842	18759	4718
11	72873	18778	4720
12	71605	18812	4725
13	68374	18759	4718
14	73385	18905	4731
15	67156	31095	4721
16	68494	34612	4726

Table 18. The values of sumI_0 of currents monitored at R3, R4 and R5 for an AG fault at different buses.

Fault Location	sumI_0 – R3	sumI_0 – R4	sumI_0 – R5
1	10.272	4.4205	17.74
2	19.58	5.375	4.9475
3	3.42	3.8354	26.49
4	15.89	4.2441	23.162
5	13.124	3.5016	26.748
6	7.1568	1.9245	22.486
7	3.972	1.0735	14.895
8	10.969	2.9396	26.874
9	24.79	0.8364	19.847
10	4.9506	1.171	25.791
11	4.5877	25.368	23.584
12	7.956	1.895	24.114
13	3.9245	0.9169	22.212
14	12.227	3.0946	27.201
15	4.7208	1.246	16.995
16	8.5112	2.2477	24.648

5. CONCLUSIONS

The distribution power system automation techniques have been widely adopted and the infrastructure of communication has been developed. The protection schemes based on microprocessors with communication capabilities are utilized, so that the status of the relays and breakers can be obtained from the distribution power system supervisory control and data acquisition system, which can serve as an information exchange platform. Based on the platform, the protection coordination and adaptation can be dealt with flexibly. In this paper, a new agentbased fault diagnosis scheme was proposed. The algorithm used the entropy calculation along with wavelet transform of current signals to classify and locate the fault in distribution network with distributed generation. The relay agents exchange data between each other through a telecommunication network. The proposed algorithm, based on the information collected by different relay agents, was able to classify and then locate the fault such that the faulted area can be correctly isolated. Simulation carried out using the CIGRE MV benchmark system and the practical 66 kV system of Alexandria showed that the proposed protection scheme is capable of classifying and locating the fault under different fault conditions, for radial and meshed networks.

REFERENCES

1 Z. Chen and W. Kong, "Protection Coordination Based on Multi-Agent for Distribution Power System with Distribution Generation Units, 2007, pp. 1-5. http://vbn.aan.dk/fbpretrieve/14217079

2 P. Barker and R. W. de Mello, "Determining the Impact of Distributed Generation on Power Systems: Part 1—Radial Power Systems," Proceedings of IEEE Power Engineering Society Summer Meeting, Seattle, July 2000, pp. 1645-1658.

3 N. Perera, A. D. Rajapakse and T. E. Buchholzer, "Isolation of Faults in Distribution Networks with Distributed Generators," IEEE Transactions on Power Delivery, Vol. 23, No. 4, October 2008, pp. 2347-2355.

4 "IEEE Standard for Interconnecting Distributed Resources with Electric Power Systems," IEEE Std., 2003.

5 A. K. Pradhan, A. Routray and S. M. Gudipalli, "Fault Direction Estimation in Radial Distribution System Using Phase Change in Sequence Current," IEEE Transactions on Power Delivery, Vol. 22, No. 4, 2007, pp. 2065-2071.

6 W. El-Khattam and S. Sidhu, "Restoration of Directional Overcurrent Relay Coordination in Distributed Generation Systems Utilizing Fault Current Limiter," IEEE Transactions On Power Delivery, Vol. 23, No. 2, 2008, pp. 576-585.

7 H. H. Zeineldin and J. L. Kirtley, "A Simple Technique for Islanding Detection with Negligible Nondetection Zone," IEEE Transactions on Power Delivery, Vol. 24, No. 2, April 2009, pp.779-786.

8 S. M. Brahma and A. A. Girgis, "Development of Adaptive Protection Scheme for Distribution Systems with High Penetration of Distributed Generation," IEEE Transactions on Power Delivery, Vol. 19, No. 1, 2004, pp. 56-63.

9 Z. Y. He, X. Q. Chen and G. M. Luo, "Wavelet Entropy Measure Definition and Its Application for Transmission Line Fault Detection and Identification (Part I: Definition and Methodology)," Proceedings of International Conference on Power System Technology, Chongqing, 2006, pp. 1-6.

10 Z. Y. He, X. Q. Chen and G. M. Luo, "Wavelet Entropy Measure Definition and Its Application for Transmission Line Fault Detection and Identification (Part III: Transmission Line Faults Transients Identification)," Proceedings of International Conference on Power System Technology, Chongqing, 2006, pp.1-6.

11 Z. M. Li, W. X. Li and R. Y. Liu, "Applications of Entropy Principles in Power System: A Survey," IEEE/PES of Transmission and Distribution Conference and Exhibition, Dalin, 5 December 2005, pp. 1-4.

12 Y. Tomita, C. Fukui, H. Kudo, J. Koda and K. Yabe, "A Cooperative Protection System with an Agent Model," IEEE Transactions on Power Delivery, Vol. 13, No. 4, 6 October 1998, pp. 1060-1066.

13 S. Sheng, K. K. Li, W. L. Chan, X. J. Zeng and X. Z. Duan, "Agent-Based Self-Healing Protection System," IEEE Transactions on Power Delivery, Vol. 21, No. 2, April 2006, pp. 610-618.

14 D. V. Coury, J. S. Thorp, K. M. Hopkinson and K. P. Birman, "An Agent-Based Current Differential Relay for Use with a Utility Intranet," IEEE Transactions on Power Delivery, Vol. 17, No. 1, 2002, pp. 47-53.

15 K. Rudion, Z. A. Styczynski, N. Hatziargyiou, S. Papathanassiou, K. Strunz, O. Ruhle, A. Orths and B. Rozel, "Development of Benchmarks for Low and Medium Voltage Distribution Networks with High Penetration of Dispersed Generation," 2007, pp.1-7. http://users.nuta.gr/stpapath/paper-2.61.pdf

Index